HUMAN GENETICS
AND
SOCIAL PROBLEMS:
A Book of Readings

Edited by
Thomas R. Mertens
Sandra K. Robinson
Ball State University

MSS Information Corporation
655 Madison Avenue, New York, N.Y. 10021

This is a custom-made book of readings prepared for the courses taught by the editors, as well as for related courses and for college and university libraries. For information about our program, please write to:

MSS INFORMATION CORPORATION
655 Madison Avenue
New York, New York 10021

MSS wishes to express its appreciation to the authors of the articles in this collection for their cooperation in making their work available in this format.

Library of Congress Cataloging in Publication Data

Mertens, Thomas Robert, 1930-   comp.
  Human genetics and social problems.

  Includes bibliographies.
  CONTENTS: Sutton, H. E. Human genetics: a survey of new developments.--Heller, J. H. Human chromosome abnormalities as related to physical & mental dysfunction.--Libassi, P. T. Getting down to business in genetics-research, diagnosis, treatment. [etc.]
  1. Human genetics--Social aspects--Addresses, essays, lectures. I. Robinson, Sandra K., joint comp. II. Title. [DNLM: 1. Genetic intervention. 2. Genetics, Human--Collected works. 3. Social problems--Collected works. QH 431 M575h 1973]
QH431.M386   301.24'3   73-3037
ISBN 0-8422-0230-7

CONTENTS

28456

Preface

The science of genetics is generally dated from 1900 when Mendel's work was "rediscovered" by Correns, DeVries, and Von Tschermak. During the past seven decades genetics has become a major unifying force in biology, providing a foundation for explaining the mechanism of evolution and the transmission of traits at the cellular and organismic levels. The last quarter of a century has seen especially dramatic contributions by biochemical and microbial geneticists to the study of molecular biology.

Progress in studying human genetics has been hampered by the very nature of man himself - the long generation time, the relatively small size of human families, and, of course, the impossibility of carrying out experimental matings. Nonetheless, progress has been made and human genetics has developed as a fascinating science in its own right. Indeed, recent years have seen such rapid progress and the future offers such bright prospects, that many scientists and laymen forsee the possibility of using genetic knowledge to manipulate human reproduction and heredity in order to correct human genetic defects, and perhaps ultimately to alter the gene pool of the human species and to direct human evolution.

Growing out of the discoveries about human genetics and the technology these discoveries will make possible are major problems having social, legal, and ethical implications. Today we are already confronted with certain social, legal, and ethical problems by such developments as genetic counseling, sperm banks, artificial insemination by donors, amniocentesis, and the abortion of genetically-defective foetuses. Tomorrow the problems could be much more complicated and involve the cloning of human eggs with the consequent asexual reproduction of a large number of genetically identical individuals, genetic surgery or gene repair (perhaps by means of viruses), legislative control of who can and cannot reproduce with enforced regulations about sterilization, and ultimately human control over the genetic future of the human species.

There are those who would argue that these problems lie in the future, that they may, in fact, never materialize, and that we can always deal with them if and when they do arise. Others suggest that these potential problems are of the same magnitude as those that confronted society with the development of the atomic bomb. Society was generally unprepared for the social, moral, and ethical problems posed by the Atomic Age. Should it also enter the Genetic Age unprepared?

Would it not be better to indulge in some thoughtful intro-
spection while we have the time in which to do it, rather
than be forced to make decisions having profound biological,
social, and ethical implications under the pressure of some
"genetic bomb"?

It is the conviction of the editors of this book that
some forethought should be given to these questions. Certainly college biology majors, and especially future high school
biology teachers, need to be informed about the developments
in human genetics and the implications of these developments
for genetic engineering.

The idea for compiling this book of readings was Mrs.
Robinson's. As an undergraduate biology major at Ball State
University, she was a student in Dr. Mertens' genetics course,
where she became interested in learning more about human
genetics. As a prospective biology teacher she saw the importance of having sound information at her command to convey to her own students. Both she and Dr. Mertens are con-
vinced that the educated citizen of the 1970s must know
something about the discoveries being made in human genetics
and the implications of these discoveries for genetic engineering.

The readings in this book have been selected from a
variety of sources - scientific journals, journals for science
educators, and from popular magazines such as Saturday Review.
Dr. Mertens judged the article from the point-of-view of the
genetics instructor. Mrs. Robinson evaluated each article
from the student's view-point. The articles included present
a controversial picture of a controversial subject. They
have been selected for their readability, understandability,
and biological significance. On occasion Mrs. Robinson exer-
cised her editorial prerogatives and rejected articles
selected for inclusion by Dr. Mertens because she did not
regard them as suitable for the intended readers.

What has resulted, then, is a book of readings we hope
the student audience will find interesting, readable, and
thought-provoking. Certainly it is controversial. Ultimately we hope it will make the readers better able to understand
and cope with some of the problems with which we are likely
to be confronted while living in the last quarter of the
twentieth century.

December, 1972   T.R.M.
Muncie, Indiana   S.K.R.

The Problem: Concerns about Genetic Engineering

The prospect of genetic engineering frightens a good many people - both laymen and scientists alike. For man to intervene in the human reproductive process, to alter the genetic constitution of an individual, and perhaps ultimately to change the genetic make-up of the entire species is, indeed, awe-inspiring. Grave social, legal, and ethical questions are raised by such genetic manipulation. Although some of the technology of genetic engineering has not yet been perfected, we would do well to consider the questions this technology raises before we are forced to make decisions under pressure.

The editors have chosen as the introductory article for this book, a paper by Caryl Rivers, a non-scientist, who discusses some of the discoveries that suggest that sophisticated genetic engineering will soon be possible. Rivers notes that even such a simple procedure as genetic counseling raises ethical questions. On the other hand, cloning, sperm banks, and genetic surgery pose enormously complex and frustrating moral questions. If there are problems created by the attempt to improve the genetic constitution of one individual - to correct a genetic error he possesses or to prevent his transmitting it to the next generation - how much more complicated are the questions raised by the possibility of controlled human reproduction with the goal of altering the evolutionary future of the entire species!

The editors intend for the article by Caryl Rivers to stimulate the interest of the reader by suggesting some of the techniques that might be used to accomplish genetic engineering and some of the problems that would result if these techniques were applied. The fact that an article of the nature of Rivers' was published in a journal such as Saturday Review suggests that the problems of genetic engineering are of concern to a wide sector of the educated public in the United States. Many people in addition to scientists and physicians are aware of the potential for both good and evil that recent advances in genetics portend.

The remainder of the book is intended to provide the student with deeper insights into the problems alluded to in Rivers' article. After looking at some of the more recent advances in human genetics and cytology in part two of the book, the reader is then introduced to some of the techniques that might be used to accomplish genetic engineering. This section of the book includes articles by Nobelists Muller, Lederberg, and Watson. The fourth section of the book examines diverse viewpoints relative to the social, ethical, and legal implications of genetic engineering and controlled reproduction. The book concludes with a selected list of references that the interested reader might use to pursue the topic of genetic engineering and its social and ethical implications to greater depth.

One final general comment should be made about the articles selected for inclusion in this book: The editors are aware that the articles are quite diverse, that they vary in both literary and scientific merit and certainly that not all important and meritorious papers have been included. They do present a collage of a topic that is of wide interest to a large number of people today.

GENETIC ENGINEERING PORTENDS A GRAVE NEW WORLD

Caryl Rivers

A society matron proudly introduces her two sons and her daughter. They are her children by every biological rule but one: She has never been pregnant. Sex cells from her body and her husband's body were implanted three times in the womb of a "proxy mother," who was paid a union wage to carry each fetus and give birth.

An astronaut is carried aboard a space vehicle destined to probe the outer limits of this galaxy. He has no legs. Legs would be only an inconvenience during the years in which the astronaut will be confined to his spaceship. For this reason he was programed to be born legless.

A Latin American dictator has some skin tissue scraped from his left arm. Nine months later, five hundred babies emerge from a factory that contains five hundred artificial wombs. The babies are genetic carbon copies of the dictator.

Although fiction today, these events could become realities with relatively slight advances in man's most adventurous and morally complex science, genetic engineering.

This new science is experimenting with a technique that would make possible the manipulation of an embryo during gestation so as to change its physical characteristics. It may offer an alternative to natural human reproduction—a process that would allow the implantation of a fertilized ovum in the womb of a "host" mother or in an artificial womb inside a laboratory. Not too far off, according to specialists in the field, is the possibility of creating children with only one parent who will be biological duplicates of that single parent. Genetic engineering, in short, is on the brink of revolutionizing the traditional concepts of man, God, and creation.

The moral questions posed by recent advances in this and related life sciences are no longer speculative. They have to be faced as practical realities, and, in parts of the scientific community, this is now being done. Indeed, a heated debate is in progress. Some scientists encourage genetic experimentation, putting their faith in man's rational power. They say, in effect, that the only thing too sacred to tamper with is scientific investigation itself. Others, however, point to numerous instances in which technology has outrun man's wisdom (nuclear stockpiles, for example), and they warn that if guidelines in genetic technology are not immediately set forth, geneticists may bring into being a terrifying, uncontrollable Huxleian nightmare.

SATURDAY REVIEW, April 8, 1972, pp. 23-27.

9

Those who urge restraint say it is past time to start thinking about controls for genetic technology. Scientific advances have already made it possible to change fundamentally the human reproductive process. The Caesarean section is now a common operation and constitutes a basic tampering with the way in which babies are born. There have already been successful blood transfusions to unborn children in cases of Rh blood incompatibility. An estimated 25,000 women whose husbands are sterile resort to artificial insemination in this country each year, and more than a third of them give birth as a result.

Test-tube fertilization of a human ovum is already a reality. In 1950 Dr. Landrum Shettles of Columbia University said he had achieved in vitro —outside the womb—fertilization of an ovum and had maintained the embryo's life for six days. In 1961 Dr. Daniele Petrucci of the University of Bologna claimed he had fertilized an egg in vitro and that the embryo lived for twenty-nine days, although it became enlarged and deformed. His work was condemned by Pope John XXIII.

Perhaps the most advanced work in the field is being done at Cambridge University in England by Drs. Robert G. Edwards and P. S. Steptoe. They have produced the best evidence to date of true in vitro fertilization and are presently attempting to implant the fertilized egg in the uterus of the donor. They obtain the ova for their experiments from volunteer women who are seeking to overcome infertility. One woman, for example, had blocked Fallopian tubes that made normal conception impossible.

Once an ovum can be successfully implanted in the womb, the way is open for "proxy mothers." A fertilized ovum could be removed from the womb of a woman after she had conceived and then be implanted in the uterus of another woman who would ultimately give birth. The child, of course, would carry the genetic identity of its true parents. The proxy mother

would be only a temporary host, with no genetic relationship to the child. Technology could go one step further and simply eliminate the process of pregnancy and childbirth altogether. A human embryo could be removed from the uterus and placed in an artificial womb. A prototype of such a device was constructed by Dr. Robert Goodlin at Stanford University. It is a pressurized steel chamber containing an oxygen-rich saline solution that would push the oxygen through the body of the fetus. A fetus put into the chamber, however, would perish from its own wastes, which the placenta draws off in natural pregnancy. Dr. Goodlin is doing no further work in the area and has announced that he does not intend to resume his experiments with the artificial womb.

At the National Heart Institute at Bethesda, Maryland, Drs. Warren Zapol and Theodore Kolobow are placing lamb fetuses in a liquid solution and attaching the umbilical cords to machinery that contains an artificial lung, a pump, and a supply of nutrients. The fetuses survive several days before "dying" of cardiac arrest.

Such ex utero gestation might offer some obvious advantages. If made common practice, doctors would be able to monitor the growth of the embryo closely. A new specialist, the "fetologist," would check to ensure normal development and catch possible medical problems in the earliest stages. The trauma of birth, from which some babies die or suffer injury, would be avoided altogether. Ex utero development would make more likely manipulation of the developing fetus to change its sex, size, and other physical characteristics.

French biologist Jean Rostand speculated that, once the size of the human brain is not limited by the size of the female pelvis, it might be possible to double the number of fetal brain cells. Would this produce superintelligent people or monstrous misfits? Where will the line be drawn between legitimate medical practices and a Frankenstein-like toying with the human condition? And who will draw the line?

Another possibility of genetic engineering that raises serious moral questions is the creation of what are called clones—carbon copies of a human being already in existence, resulting from a process much more revolutionary than the merging of sperm and egg in a test tube.

Normally, the sexual process by which life begins guarantees the diversity of the species. The child inherits characteristics from both the mother and the father so that, while he may resemble either or both of them, he is unique. There are two kinds of cells in the human body—sex cells (ova and sperms) and body cells. Each body cell has a nucleus containing forty-six characteristic-determining chromosomes. Sex cells have only half that number. A merger of two sex cells is required for the resulting zygote to have a complete set of chromosomes and to begin dividing.

But human reproduction could take another route. Body cells contain the requisite number of chromosomes to start a new individual, but they cannot divide the way the fertilized egg does. The body cells have very specialized jobs to do. Some go into making hair, others teeth, others skin, and so forth. Only the part of the body-cell mechanism that is useful in its specific job gets "switched on." If a body cell could receive a signal to switch on all of its mechanisms, it would divide and subdivide and multiply, finally developing into a new human being.

Accordingly, if such a "switching on" technique were developed, a body cell could be taken from a donor—scraped from his arm, perhaps--and be chemically induced to start dividing. The cell could be implanted in an artificial womb or in the uterus of a woman, where, presumably, it would develop like a normal fetus. The baby would be genetically identical to the donor of the cell—his twin, a generation removed. It would have only one true parent. Its "mother" would be, like the proxy mother, only a temporary host.

The possibilities of human cloning are enough to startle even a science-fiction writer. Societies would be tempted to clone their best scientists, soldiers, and statesmen. An Einstein might become immortal through his carbon copies. Or a Hitler. What better way for a dictator to extend his power beyond the grave? A carbon copy might be the ultimate expression of human egoism.

Clones have already been produced from carrots and frogs. Dr. J. B. Gurdon of Oxford University produced a clone from an African clawed frog by taking an unfertilized egg cell from a frog and destroying its nucleus by radiation. He then replaced it with the nucleus of a cell from the intestines of a tadpole. The egg began to divide as if it had been normally fertilized. The result: a twin of the donor tadpole.

Some scientists estimate that human cloning will be possible by the end of the 1970s. If human cloning were ever practiced on a wide scale, it would drastically affect the course of human evolution. Nobel Prize-winning geneticist Dr. Joshua Lederberg has said he can imagine a time in the future when "clonishness" might replace the present dominant patterns of nationalism and racism.

Cloning is not the only avenue man might take to redesign himself. Advances in molecular biology could lead to genetic surgery: the addition of genes to or removal of them from human cells. In experiments with mouse cells, scientists added healthy new genes to a defective cell. The new genes not only corrected an enzyme deficiency in the cell but were duplicated when healthy new cells began to divide. It is now possible, in principle, to remove defective cells from a human patient, to introduce new genes, and to place the "cured" cells back into the patient.

Genetic surgery might also be put to more ambitious uses. J. B. Haldane, the late, renowned British geneticist, was one of the first to speculate and predict that men would tailor-make new kinds of people for space travel. He suggested the legless man referred to earlier, who would be ideally suited to living in the cramped quarters of a space capsule for long journeys; men with prehensile feet and tails for life on asteroids where low gravity makes balance difficult; and muscular dwarfs to function in the strong gravitational pull of Jupiter.

Genetic engineering might also bring to life one of the mixtures of man and animal known in all the myths of man: the Minotaur, the Centaur, and the Gorgons. So far, however, human and animal chromosomes have been successfully mingled only in tissue-culture experiments.

Modern medicine may be able to produce not only man-animal combinations but a man-machine combination for which a name has already been coined: cyborg. We are familiar with the use of artificial material in the human body: plastic arteries, synthetics to replace damaged bone, or even an artificial heart. Who will draw the dividing line between man and robot?

There are few existing guidelines for the manipulation of "living" material and human beings. Reports from Britain describe experiments in which pregnant women awaiting abortion have been exposed to certain kinds of sound waves to determine if there will be extensive chromosome damage in the fetus. Is it immoral to damage a fetus for experimental purposes, even though it will be aborted? Does an embryo "grown" in a laboratory have any legal rights?

Even the practice of genetic counseling—guidance based upon an individual's chromosomal make-up—which is becoming increasingly common, raises ethical questions. Would an individual who is marked as a carrier of deleterious genes carry a social stigma as well? Dr. Marc Lappé of the Institute of Society, Ethics, and the Life Sciences, a group set up to examine the moral questions posed by

science, thinks he would.

"It could lead to a subtle shift in the way we identify people as human," he says. "We could say, 'Oh, he has an extra chromosome.' We could then identify him as being qualitatively different. He is somewhat imperfect, perhaps not as human as the rest of us."

Suppose, for example, a genetic counselor knows that a newborn baby has an extra Y chromosome. There is evidence that the extra Y chromosome predisposes an individual to aggressive behavior. Should that become part of his medical record? What impact might that knowledge have on a school principal if a child with an extra Y becomes involved in typical childish pranks? Should the parents know, or would the knowledge have an adverse effect on the way they bring up the child?

The agonizing moral question posed by almost all aspects of genetic engineering has divided the scientific community. Hearing the arguments among scientists, one is reminded of the story about two men in a jail cell. One looked out and saw the mud; the other saw the stars.

Dr. Robert Sinsheimer of the California Institute of Technology sees the promise: "For the first time in all time a living creature understands its origin and can undertake to design its future. Even in the ancient myths man was constrained by his essence. He could not rise above his nature to chart his destiny."

Dr. Sinsheimer goes on to say that those who oppose genetic engineering "aren't among the losers in the chromosomal lottery that so firmly channels our human destiny." Repugnance to advances in genetic technology "isn't the response of four million Americans born with diabetes, or the two hundred fifty thousand children born in the United States every year with genetic diseases, or the fifty million Americans whose I.Q.s are below ninety."

Dr. Salvador Luria of MIT, a Nobel laureate, takes the opposite viewpoint. "We must not ignore the possibility that genetic means of controlling human heredity will become a massive means of human degradation. Huxley's nightmarish society might be achieved by genetic surgery rather than by conditioning and in a more terrifying way, since the process would be hereditary and irreversible."

The work of Steptoe and Edwards in Cambridge has been extremely controversial among scientists. Is implantation of a fertilized ovum back into the uterus a desirable medical advance or the opening of a Pandora's box? Dr. Joseph Fletcher of the University of Virginia Medical School argued the former in a recent article that he published in the *New England Journal of Medicine.*

"It seems to me," he wrote, "that laboratory reproduction is radically human compared to conception by ordinary heterosexual intercourse. It is willed, chosen, purposed, and controlled, and surely these are among the traits that distinguish Homo sapiens from others in the animal [world], from the primates down. Coital reproduction is therefore less human than laboratory reproduction; more fun, to be sure, but with our separation of baby-making from love-making, both become more human, because they are matters of choice, not chance. I cannot see how humanity or morality [is] served by genetic roulette."

At the opposite pole, Dr. Leon Kass, the executive director of a committee on the life sciences and social policy for the National Academy of Sciences, argues that the further reproduction is pushed into the laboratory the more it becomes sheer manufacture.

"One can purchase quality control of the product only by the depersonalization of the process," he says, going on to ask, "Is there not some wisdom in the mystery of nature that joins the pleasure of sex, the communication of love, and the desire for children in the very activity by which we continue the chain of human existence? Is not human procreation, if properly understood and practiced, itself a humaniz-

ing experience?"

Dr. Kass points to the lonely depersonalized experiences old age and dying have become as a result of medical technology. The aged are kept alive but barely able to function, and they die in institutions surrounded by clacking machinery and uncaring strangers. He also believes that laboratory reproduction will deal a near fatal blow to the human family.

"The family is rapidly becoming the only institution in an increasingly impersonal world where each person is loved not for what he does, or makes, but simply because he is. Destruction of the family unit would throw us, even more than we are now, on the mercy of an impersonal, lonely present."

There are some members of the scientific community who take what might be called an "amber light" approach to genetic technology: Proceed with extreme caution.

Isaac Asimov, the biochemist who is also well known as a science-fiction writer, says we should be very sure what we're about before we start tinkering with genes. "We should intervene if we have reasonable suspicion that we can do so wisely. If our choice is doing nothing or doing something without knowing, we should do nothing. But wise change is better than no change at all. It's the same as changing the environment. Every time we build a dam there is a gain and there is a loss. What has usually happened is that we have considered the short-term gain without thinking what the long-term effect will be."

Dr. George Wald, a Nobel laureate at Harvard who has spoken out on a wide range of social issues, emphasizes the fact that every organism alive today represents an unbroken chain of life that stretches back some three billion years. That knowledge, he says, calls for some restraint. The danger he senses in genetic technology is a movement toward reducing man's unpredictability. "With animals we have abandoned natural selection for the technological process of artificial se-

13

lection. We breed animals for what we want them to be: the pigs to be fat, the cows to give lots of milk, work horses to be heavy and strong, and all of them to be stupid. This is the process by which we have made all of our domestic animals. Applied to men, it could yield domesticated men."

Wald sees the trends of modern life working to erode one of the glories of being human: free will. "Free will is often inefficient, often inconvenient, and always undependable. That is the character of freedom. We value it in men. We disparage it in machines and domestic animals. Our technology has given us dependable machines and livestock. We shall now have to choose whether to turn it to giving us more reliable, efficient, and convenient men, at the cost of our freedom. We had better decide now, for we are already not as free as we once were, and we can lose piecemeal from within what we would be quick to defend in a frank attack from without."

Dr. Harold P. Green, a law professor at George Washington University, argues for the creation of an agency within the government whose sole task would be to make the case against technological advances. With any technology, he says, the benefits are immediate and obvious, and there are powerful vested interests to articulate them. The risks are more remote, and there are few people to argue these points. A prestigious new agency so mandated might right the present imbalance.

Senator Walter Mondale has proposed a fifteen-man presidential advisory committee that would spend two years looking into the moral problems posed by the life sciences. The commission would devise a structure to give society controls over such things as genetic technology. The proposal was approved by the Senate and, at last report, is languishing in a House committee.

Scientists themselves have shown growing awareness of the moral implications of their work. "Scientists have always had a supreme obligation to be concerned about the uses of their work," Isaac Asimov says. "In the past an ivory-tower scientist was just stupid. Today he's stupid and criminal."

But is it enough that scientists develop sensitivity to the moral complexities of the new technology? Can they alone grapple with choices so fundamental they could affect the future of human heredity? James Watson has called the idea that science must always move bravely forward "a form of laissez-faire nonsense dismally reminiscent of the creed that American business, if left to itself, would solve everybody's problems."

Seeing a child lying crippled by a genetic disease, one can't help thinking that, if such a scourge could be lifted from the children of the future, it would be worth the risk of any Brave New Worlds. It is probably unwise and perhaps impossible to barricade any street of scientific inquiry and say: No Admittance. But neither can we blind ourselves to the consequence of traveling that street until it is too late. Dr. Kass has summarized the question posed by genetic engineering: "Human heredity is intricate and mysterious. We must face the prospect of intervention with awe, humility, and caution. We may not know what the devil we are doing."

The Background: Progress in Human Genetics

The possibility of having genetic engineering become a
reality in our lifetime is ever increasing. The technology
necessary for genetic engineering is dependent on basic research discoveries in genetics and cytology - including human
genetics and cytology. Especially noteworthy advances have
been made in the study of human chromosomes - beginning with
the discovery by Tjio and Levan in 1956 that the human diploid
chromosome number is 46 rather than 48 as had been generally
accepted for almost 35 years. The papers in this section of
the book document recent advances in human genetics and cytology and afford a glimpse at anticipated developments in
those fields.

Discoveries in biochemical, molecular, and immunogenetics
are mentioned by Sutton in the first of these papers. Briefly discussed are some of the many genetic variants of human
hemoglobin and how these might arise by single nucleotide
substitutions in the DNA molecule. Also briefly introduced
are the metabolic disturbances, phenylketonuria (PKU) and
glucose-6-phosphate dehydrogenase (G6PD) deficiency.

The paper by Heller discusses discoveries about human
chromosomes and cytogenetics made during the last fifteen
years and includes a discussion of the 47,XYY chromosome
complement and the social, legal, and ethical questions
raised by the purported aggressive and criminal tendencies of
XYY individuals. The paper by Heller includes an extensive
reference section which will be useful to the interested
student.

The final selection in this section of the book will
give the reader some idea of the directions in which human
genetics is likely to be moving in the next decade. The
article describes the establishment with federal government
support of seven major research centers, each designed to
study specific kinds of human genetic defects. The need for
these centers is explained and a brief account is given of
some of the expectations that are held for them and for
progress in human genetics.

Human Genetics:
A Survey of New Developments

H. ELDON SUTTON

IT IS almost exactly ten years ago that human cytogenetics became a distinct discipline with emphasis on
the relationship of chromosome constitution to human
variation. Prior to that, efforts had been devoted primarily
to development of techniques for chromosome study. By
1957, several procedures had become available, enabling
direct examination and characterization of somatic chromosomes of most tissues. Foremost among these procedures were (1) the ability to culture cells *in vitro* at
least for a short period (several cell divisions); (2) the
use of colchicine to arrest cell division in metaphase, leading to the accumulation of many cells at the optimum
stage for study; and (3) treatment of cells with buffers of
low osmotic strength, resulting in dispersion of the chromosomes with much less overlapping.

Attention was dramatically focused on the new field of
human cytogenetics by the report that human beings have
only 46 chromosomes rather than the 48 that had been
the accepted number for almost 35 years. These 46 consist of 22 homologous pairs, the autosomes, plus a pair of
sex chromosomes (two X chromosomes in females and an
X and Y chromosome in males). Chromosomes differ in
length and in position of the centromere (or kinetochore),
the point at which spindle fibers attach in division and
which play a major role in assuring correct distribution of
daughter chromosomes to new daughter cells. In somatic
cell division (mitosis), colchicine halts chromosome division after the new chromosome arms have separated but
before the centromere has divided. Typically, chromosomes prepared in this manner appear as variations on the
letter X, the centromere being the point at which the four

THE SCIENCE TEACHER, 1967, Vol. 55, No. 15, pp. 51-55.

16

arms meet. (Figure 1) Its position may be anywhere along the chromosome, although in man there are no chromosomes with centromeres on the end. By the size of the chromosomes and position of the centromere, human chromosomes can be classified into seven groups, usually designated by letters A through G. Some specific chromosome pairs can be recognized consistently within these groups; others cannot. With rare exception, the chromosome complements (karyotypes) of normal persons are indistinguishable except for the XX or XY combination.

The announcement in 1959 that persons with mongolian idiocy have 47 rather than 46 chromosomes stimulated an enormous effort to examine chromosomes of many other disorders of obscure etiology. In mongolism (also called Down's syndrome), there are three rather than two of one of the G group chromosomes, often designated chromosome 21. This condition is therefore said to result from *trisomy* of a G chromosome and is called the trisomy 21 syndrome by some. The features of the condition, which include mental deficiency, a typical facial appearance, and characteristic palm prints, can be attributed to the unbalancing effect of three copies of many genes as opposed to the normal two copies.

Only two other autosomal trisomy syndromes are known, one involving a group D (13-15) chromosome, the other involving a group E (No. 18). Both are asso-

Figure 1. *Chromosome spread and karyotype of a normal human male.*

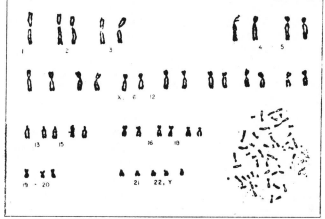

17

ciated with multiple malformations, and most affected infants do not survive more than a few weeks.

Abnormal numbers of sex chromosomes occur approximately once in 500 births. The most common is the combination XXY, which results in a male with Klinefelter's syndrome. Such persons may be essentially normal in appearance, although most often they have features which are somewhat eunochoid, and they are infertile. The most common defect in females is Turner's syndrome, or gonadal dysgenesis, rarer than Klinefelter's syndrome, with only 45 chromosomes including one X and no Y chromosome. The karyotype of such a person is written 44A + XO, the letter O indicating absence of the homologous chromosome.

The recognition that in human beings the XXY complement produces males and the XO females was an indication that sex is determined by presence or absence of a Y chromosome. This is in contrast to *Drosophila,* where the ratio of X chromosomes to autosomes determines the sex. Confirmation came with the discovery of cases with greater numbers of X chromosomes—XXXY and XXXXY. These also are males, very little different from XXY or, for that matter, XY males. The XYY karyotype has been observed a number of times and often results in increased height and aggressive behavior; otherwise, the person is essentially normal.

ACCIDENTS occasionally give rise to chromosomes of abnormal structure. Of particular importance are deletions and translocations. Deletions are losses of a chromosome segment and often arise when chromosomes break and fail to rejoin. Only those parts attached to a centromere are properly distributed in cell division, and segments distal to a chromosomal break are lost. The result is partial monosomy for the lost segment, with consequent genic imbalance and expression of recessive genes on the corresponding segment of the normal homologous chromosome.

Most major deletions are incompatible with life. Several have occurred a number of times, of which the *cri-du-chat* or "cat's cry" syndrome is the best known. This results from deletion of part of the short arm of chromosome 5, and affected persons are severely retarded mentally, with a characteristic facial appearance and often a peculiar cat-like wail when young. Two other deletions seen in a num-

18

Figure 2. Translocation between two nonhomologous chromosomes showing pairing in meiosis and the six kinds of gametes theoretically possible. Two pairs of chromosomes are shown. Other chromosomes in the complement would behave normally. Crossing over in the translocated segment would produce additional combinations. Although six possibilities are shown, these need not be equally likely to occur.

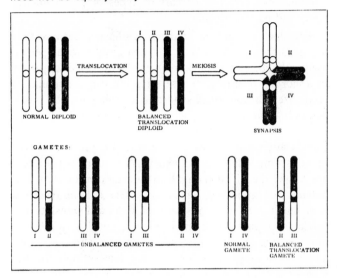

ber of patients involve partial deletion of the short arm of chromosome 18 and partial deletion of its long arm.

One form of malignancy, chronic granulocytic leukemia, is due to partial deletion of the long arm of a G group chromosome, thought to be the same that is trisomic in mongolism. Such patients start out with a normal complement, but a deletion occurs in the marrow cells, giving rise to a clone of abnormal cells responsible for this form of leukemia. Other forms of malignancy, including the other leukemias, have not been associated with chromosomal abnormalities.

If breaks occur in two nonhomologous chromosomes, exchange of distal segments sometimes occurs during the repair process, and translocation is said to have occurred. The cell in which this happens still has a normal gene complement, but the genes are now associated into abnormal combinations. The cell and those that arise from it by simple mitotic division usually are normal, but the reduction division of meiosis can no longer occur normally. An illustration of what may happen is given in Figure 2.

19

Gametes from such a meiotic division may be normal, a balanced translocation, or an unbalanced combination. The balanced translocation, combined with a normal gamete, produces a zygote similar in constitution to the cell in which the translocation arose. The unbalanced combinations give rise to various types of partial monosomy and partial trisomy.

Translocations are of medical significance because they are a mechanism by which an apparently normal person (with a balanced translocation) produces a high frequency of abnormal offspring. These offspring may be similar to recognized chromosomal syndromes, or they may be classified only as miscarriages, being incompatible with development to live birth. Offspring who inherit the balanced translocation will be normal but will also produce abnormal offspring with high frequency. A number of pedigrees are known in which mongolism due to translocation imbalance has appeared repeatedly. The risk of defect is very much greater in offspring of balanced translocation carriers compared to normal persons, and it is important in estimating risk for future offspring to ascertain whether a child with mongolism, for example, has a simple trisomy with 47 chromosomes or an unbalanced translocation.

NOWHERE has human genetics made a greater contribution to general genetic understanding than in the area of biochemistry and biochemical genetics. Much of this understanding comes from studies of inherited variants of hemoglobin. Hemoglobin, like other proteins, consists in part of long chains of amino acids attached by peptide bonds. There are 20 amino acids commonly found in proteins, and the specific sequence determines the properties of the molecule. A complete hemoglobin molecule has four such polypeptide chains. Two of the chains are of one type, designated α chains; the other two are of a second type, designated β chains. There are 141 amino acids in each α chain and 146 in each β chain.

Some 60 genetic variants of human hemoglobin are known, most of them differing from the usual form because of a slight difference in charge and hence in electrophoretic mobility. The nature of the structural variations associated with different genes was established by Ingram in 1956 working with sickle cell hemoglobin (Hb S). This form of hemoglobin leads to serious disease if the person has two genes for it. Ingram showed that the

Table 1. Amino acid substitutions in inherited human hemoglobin variants. A number of additional variants have been studied but are not given in the table.

Amino acid residue No.	Residue in Hb A	Variant	
		Substitution	Designation of Hb
α chain			
1	val [a]		
2	leu		
16	lys	glu	I
30	glu	gln	G Honolulu
57	gly	asp	Norfolk
58	his	tyr	M Boston
68	asn	lys	G Philadelphia
87	his	tyr	M Iwate
116	glu	lys	O Indonesia
141	arg		
β chain			
1	val		
2	his		
3	leu		
6	glu	{ val	S
		{ lys	C
7	glu	gly	G San Jose
26	glu	lys	E
63	his	tyr	M Saskatoon
79	asp	asn	G Accra
121	glu	lys	O Arabia
146	his		

[a] Abbreviations of amino acids are as follows: arg (arginine), asn (asparagine), asp (aspartic acid), gln (glutamine), glu (glutamic acid), gly (glycine), his (histidine), leu (leucine), lys (lysine), tyr (tyrosine), and val (valine).

only difference between Hb S and the normal Hb A is in the amino acid in the sixth position of the β chain. In Hb A it is glutamic acid; in Hb S it is valine. Studies on some three dozen of the other variants have confirmed that each varies in a single amino acid at some specific point along the chain. (Table 1) From these studies, it is clear that one important action of genes is to determine the sequence of amino acids in polypeptide chains. Different forms of a gene produce different but closely related sequences, ordinarily differing only at a single amino acid position.

The mechanisms by which genes direct protein structure have been worked out largely on other organisms, but hemoglobin variants follow the predicted pattern. Genetic information is stored in the sequence of nucleotides in deoxyribonucleic acid (DNA). A sequence of three nucleotides constitutes a *codon,* i.e., the coding unit for one amino acid. Any of four nucleotides is possible at a given position in the DNA; hence there are 64 possible nucleotide sequences in a nucleotide triplet. This is more than the

Table 2. RNA code for amino acids. The RNA code is that which functions for messenger RNA and is complementary to the DNA code. Several lines of evidence have been used to establish the code, which appears to be universal biologically. Abbreviations for acids: U (uridylic), C (cytidylic), A (adenylic), G (guanylic).

Phenylalanine	Serine	Tyrosine	Glutamic acid
UUU	UCU	UAU	GAA
UUC	UCC	UAC	GAG
Leucine	UCA		
	UCG	*Blank* [a]	*Cysteine*
UUA	AGU	UAA	UGU
UUG	AGC	UAG	UGC
CUU			
CUC	*Proline*	*Histidine*	*Tryptophan*
CUA	CCU	CAU	UGA(?)
CUG	CCC	CAC	UGG
Isoleucine	CCA	*Glutamine*	*Arginine*
	CCG	CAA	CGU
AUU		CAG	CGC
AUC	*Threonine*		CGA
AUA	ACU	*Asparagine*	CGG
	ACC	AAU	AGA
Methionine	ACA	AAC	AGG
AUG	ACG		
Valine	*Alanine*	*Lysine*	*Glycine*
GUU	GCU	AAA	GGU
GUC	GCC	AAG	GGC
GUA	GCA	*Aspartic acid*	GGA
GUG	GCG	GAU	GGG
		GAC	

[a] These two codons are blank in the sense that growth of the polypeptide chain stops at the point at which they occur and the polypeptide chain is released as "completed." They may function therefore as punctuation.

20 amino acids incorporated into proteins, and it is now established that several codons may code for the same amino acid. (Table 2) The 141 amino acids of the hemoglobin a chain would correspond to a sequence of 141 nucleotide codons, which in turn would be $3 \times 141 = 423$ nucleotides.

Most mutations involve single codon changes. By comparison of the codons for glutamic acid against those of valine, it is seen that a single nucleotide exchange could give rise to the hemoglobin S mutation, e.g., GAA→GUA. Similar results are obtained with other hemoglobin variants. By contrast, many theoretically possible amino acid substitutions could arise only by changing two of the three codon nucleotides. Such variants are rare. These results confirm that the genetic code, worked out principally with microorganisms, is the same for man and presumably for all biological systems.

Some types of changes in the DNA result in interruption of the amino acid sequence so that a complete polypeptide chain cannot be formed. Rather than a slightly altered protein product, no recognizable product is formed. Such mutants, when homozygous, lack the function associated with the normal protein whose synthesis is directed by that gene locus.

Several types of changes in the DNA may give rise to loss of recognizable protein product. If a segment of DNA is deleted, a complete polypeptide chain cannot be formed, although part of the chain corresponding to the remaining nucleotides might be formed. Ordinarily this portion would not be functional. Deletion of a single nucleotide causes a shift in the "reading frame." Translation of the nucleic acid code proceeds from one end, with three nucleotides at a time being interpreted for the corresponding amino acid. Loss of a nucleotide causes the remaining nucleotides to shift by one position, creating a new sequence of codons subsequent to the deletion. This produces a polypeptide chain which is normal up to a certain point but totally different beyond that.

Two of the 64 possible nucleotide triplets appear to be "punctuation." That is, when they occur, translation into protein stops and the polypeptide is released as "finished." These triplets, UAA and UAG in the ribonucleic acid code, can arise from other codons by single nucleotide substitution in the DNA. If they should arise in the middle of the nucleic acid message, synthesis would be interrupted at that point and incomplete chains would be released.

SO FAR we have said nothing concerning the quantitative control of gene action. Unfortunately, very little is known about this subject in higher organisms. It is clear that some genes work only in some tissues and not in others. Hemoglobin synthesis is an example. In addition, there appear to be mechanisms for informing a gene when it has done enough or when it needs to speed up. The models which have been worked out in bacteria may be valid also for other organisms.

In the special case of human hemoglobin, the rate of synthesis depends in part on the structure. Hemoglobin S is synthesized only about half as fast as normal hemoglobin, and some variants are synthesized at still slower relative rates. It is possible that the main advantage of many molecular forms is due to the rapidity with which they can be synthesized rather than to functional superiority of the completed molecule. The main disadvantage

23

of some hemoglobin variants seems to be a low rate of synthesis resulting in anemia, since the hemoglobins function quite normally in oxygen transport.

Although much of this discussion is in terms of abnormal variation, it is increasingly apparent that many human proteins, perhaps the majority, exist in a variety of inherited forms with normal function. To say that the function is normal is not to imply absence of functional differences. Perhaps the situations which might expose the functional differences do not occur in nature or occur only rarely. On the other hand, normal persons do differ in many ways from each other, and ultimately many of these differences are due to variations in gene complement. Some examples are considered in the section on pharmacogenetics.

T HE PREVIOUS section has considered ways in which mutant genes affect proteins. In terms of the physiological consequences, most mutant genes result in diminished or total loss of function of a specific protein. Since most proteins are enzymes which catalyze specific reactions of metabolism, mutations can usually be related to their effect on a certain metabolic reaction.

One of the better known examples of an inherited metabolic defect is the disease phenylketonuria (PKU). This is a typical recessive trait, in that persons with one PKU gene and a normal gene show no ill effects. A person with two PKU genes is nearly always severely mentally defective. The name phenylketonuria comes from the excretion of phenylpyruvic acid (a phenyl ketone) in the urine. The metabolic reactions important in phenylketonuria are shown in Figure 3. Normally, the amino acid phenylalanine is converted to tyrosine, which is oxidized further to recover the energy in the molecule. There are several other reactions of tyrosine of great importance physiologically, although only a small portion of the tyrosine is utilized in these pathways.

The conversion of phenylalanine to tyrosine is an oxidative step catalyzed by the liver enzyme phenylalanine hydroxylase. Phenylketonurics lack this enzyme function and therefore cannot make the conversion. The phenylalanine consumed as protein accumulates, and affected persons may have 50 to 100 times normal blood levels. Because of these high levels, side reactions occur which, under ordinary circumstances, are quantitatively unimportant.

24

Figure 3. Metabolic pathways important in phenylalanine metabolism. The reactions indicated with broken arrows ordinarily are not important quantitatively. In phenylketonuria (PKU), the normal conversion of phenylalanine to tryosine is blocked, and much of the phenylalanine undergoes reactions indicated by the broken arrows.

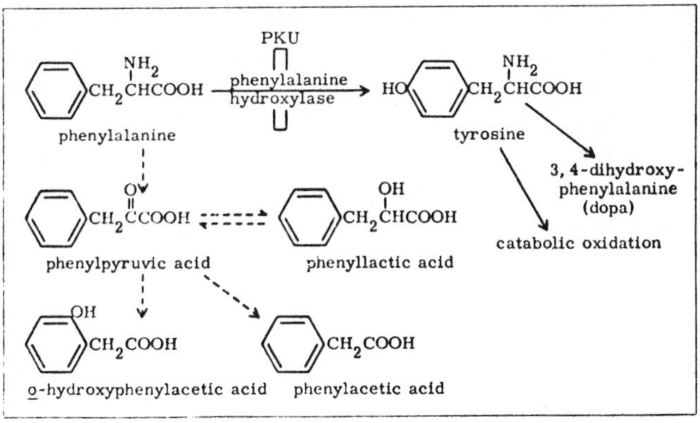

PKU

phenylalanine hydroxylase

$CH_2CHCOOH$ (with NH_2) — phenylalanine

HO— $CH_2CHCOOH$ (with NH_2) — tyrosine

3, 4-dihydroxy-phenylalanine (dopa)

catabolic oxidation

CH_2CCOOH (with O) — phenylpyruvic acid

$CH_2CHCOOH$ (with OH) — phenyllactic acid

CH_2COOH (with OH) — o-hydroxyphenylacetic acid

CH_2COOH — phenylacetic acid

These are responsible for the formation of phenylpyruvic acid, o-hydroxyphenylacetic acid, and other substances characteristic of PKU. Some of the products of these side reactions are toxic to the central nervous system and produce irreversible brain damage. It has not been established whether lack of phenylalanine hydroxylase activity results from loss of the protein or from an altered form of the protein which cannot function.

Phenylketonuria also serves as a model for one approach to treatment of inherited defects. Human beings cannot make phenylalanine, relying on dietary protein as a source. It follows that if phenylalanine were removed from the diet, persons with PKU should not be affected. The difficulty is that phenylalanine is required as a building block for body proteins, so that complete elimination would be very detrimental. The amount consumed can be controlled, however, and good results have been obtained in preventing mental defect when proper diets are instituted within the first weeks of life. Delay in starting diet control leads to irreversible changes in the nervous system, hence the institution of large-scale screening programs for detection of PKU in newborns.

Dozens of diseases involving metabolic blocks have been identified in recent years. Each varies in profoundness of the defect and means of therapy, depending on the unique metabolic and physiological aspects of the block. Some are relatively benign, and others may be within the range of what is considered normal variation.

THE TERM pharmacogenetics is applied to inherited variation which is exposed only when the person challenges his metabolism with drugs. The best known example is glucose-6-phosphate dehydrogenase (G6PD) deficiency. This enzyme is widespread in tissues and is important in glucose metabolism. Some persons inherit a form of the enzyme which is less stable than the common form. This is particularly easy to observe in red cells, where synthesis of proteins is completed early in the maturation of the cell and there is no mechanism for replenishment.

Persons with this defect are entirely normal except when challenged with certain drugs or when they eat fava beans. The drugs include the anti-malarial primaquine, which first led to recognition of the condition, as well as a variety of more common substances such as naphthalene mothballs. When exposed to these substances, G6PD-deficient red cells are unable to cope with the chemical stress and break open. The release of hemoglobin into the plasma may have very undesirable consequences, including death.

The genetic locus for G6PD is on the X chromosome. Males, with one X chromosome, are either clearly affected or nonaffected. Females may have one or two G6PD deficiency genes and show intermediate effects.

Another drug handled differently by different persons is isoniazid, used primarily in treatment of tuberculosis. Some persons inactivate the drug rapidly, others slowly, depending on the gene combinations at a specific locus.

An important area, still largely unexplored, concerns inherited variations in resistance to infection. Work with experimental animals indicates clearly the innate differences among inbred strains in ability to cope with infections. The most suggestive studies in human beings concern leprosy, where only a portion of the population appears to be susceptible. Until such differences can be tied to specific genes, the interpretation will be suspect. Nonetheless, this is an area in which significant advances can be anticipated. It also emphasizes the interrelations of heredity and environment, both always contributing to the final makeup of the individual.

HUMAN CHROMOSOME ABNORMALITIES AS RELATED TO PHYSICAL AND MENTAL DYSFUNCTION

John H. Heller

THE relationship of human disease syndromes to chromosome aberrations is assuming an increasingly greater role in the detection, diagnosis, treatment and prediction of mental and physical defects in man. By means of karyotype analysis one is enabled to recognize previously unknown syndromes and to differentiate between separate but phenotypically similar entities. Proper diagnosis permits suitable therapeutic measures to be undertaken and enables genetic counselors to assess correct risks in many instances. Recent refinements in sampling embryonic cells by amniocentesis make it feasible to determine, in high risk cases, whether the embryo has a chromosome abnormality or whether it is a male, which has a high risk of sex-linked genetic defect. Termination of pregnancy can be recommended on the basis of this knowledge.

Classes of Chromosome Abnormalities

Chromosome abnormalities have been known in plant and animal species for a very long time. They

Dr. Heller is president of the New England Institute, Ridgefield, Connecticut 06877. This paper was prepared in collaboration with Dr. George H. Mickey, the New England Institute, and was presented on August 19, 1969, as the Sixth Wilhelmine E. Key lecture at the annual meetings of the American Institute of Biological Sciences, University of Vermont, Burlington. The key lecture was established by the American Genetic Association through funds bequeathed to the Association by Dr. Wilhelmine E. Key for the support of lectures in human genetics.

THE JOURNAL OF HEREDITY, 1969, Vol. 60, pp. 239-248.

occur firstly as variations in the number of chromosomes per cell deviating from the normal two sets (maternal and paternal), existing either as complete multiples of sets, a condition called polyploidy (triploidy, tetraploidy, etc.), or as addition or loss of chromosomes within a set, a situation known as aneuploidy (monosomy, trisomy, tetrasomy, etc.). The origin of deviations in chromosome number is known to be through nondisjunction, either during the meiotic divisions in the maturation of the germ cells or during mitotic divisions in the developing individual, or through lagging of chromosomes at anaphase of cell division.

Secondly, chromosome aberrations occur as structural modifications such as duplications, deficiencies, translocations, inversions, isochromosomes, ring chromosomes, etc. These aberrations result from chromosome breakage and reunion in various patterns different from the normal sequence of loci. In most cases, especially the "spontaneous" instances, the cause of chromosome breaks is unknown, but many extraneous agents have been demonstrated experimentally to be efficacious in inducing fragmentation. Foremost among these agents is ionizing radiation but many chemical substances (alkylating agents, nitroso-compounds, antibiotics, DNA precursors, etc.) and viruses have been implicated.

Genetic Effects of Chromosome Aberrations

The striking genetic alterations accompanying chromosome aberrations were brilliantly analyzed by Blakeslee and coworkers on *Datura*, and by the *Drosophila* workers (Morgan, Bridges, Muller, Sturtevant, Painter, Patterson and many others). The task was greatly facilitated in *Drosophila* by the fortunate circumstance in the larval salivary glands where the giant polytene chromosomes exhibit intimate somatic pairing as well as characteristic banding patterns that permit identification of specific gene loci.

Particularly illuminating were Bridges' analyses of sex chromosomes and sex determination in *Drosophila*, utilizing the phenomenon of nondisjuction of the sex chromosomes and culminating in the genic balance theory of sex determination. In this insect the female normally has two X chromosomes plus the autosomes, and the male has one X and one Y. Two X chromosomes and one Y chromosome results in a female, whereas a chromosome constitution of XO produces a sterile male.

In contrast, the Y chromosome in mammals has a strongly masculinizing influence. The presence of a

28

single Y is sufficient to induce differentiation into a male phenotype in the presence of one to five X chromosomes. The XO constitution differentiates into a female phenotype in both mouse and man.

Mammalian Chromosome Studies

The first reported instance of chromosome aberration in mammals was discovered by genetic methods in the waltzing mouse by William H. Gates in 1927[37], and analyzed cytologically by T. S. Painter[68]. Many difficulties in techniques prevented accurate counting and analysis of mammalian chromosomes—large number and relative small size of chromosomes, tendency to clump on fixation, cutting of chromosomes in sectioned material, etc. Even the somatic chromosome number in man was accepted erroneously as 48 until 1956 when Tjio and Levan[88] established the correct count of 46. This count was quickly confirmed by Ford and Hammerton[28], and in 1959 the first positive correlation of a chromosome abnormality and human disease syndrome was made by Lejeune et al.[54] (also Jacobs et al.[44])—the trisomic number 21 chromosome, and Down's syndrome or mongolism. Shortly thereafter Klinefelter's[46] and Turner's[30] syndromes were identified with XXY and XO sex chromosome constitutions respectively, and in rapid succession reports of many other human chromosome abnormalities appeared, such as trisomy 17, trisomy 18, partial trisomy, ring X chromosome, sex chromosome mosaics, cri-du-chat syndrome, et cetera[9, 26].

This sudden explosion of human chromosome studies, in contrast to the long delay of confirmation in human cells of chromosome abnormalities long known in plants and other animals, was made possible by new techniques of preparation. The accumulation of many cells in the metaphase stage of mitosis with colchicine, the use of hypotonic solution to swell the cells and separate chromosomes on the spindle, the discovery that phytohemagglutinin stimulates mammalian peripheral lymphocytes to undergo mitosis, and the method of squashing or spreading on slides of loose cells taken from bone marrow or tissue culture, all contributed to the rapid and accurate analysis of mammalian and human chromosome number and structure.

Karyotype analysis involves the careful comparison of chromosomes in a particular individual to the standard pattern for human cells, including precise measurements of lengths, arm ratios and other morphological features. Special attention is given to comparison of homologous chromosomes

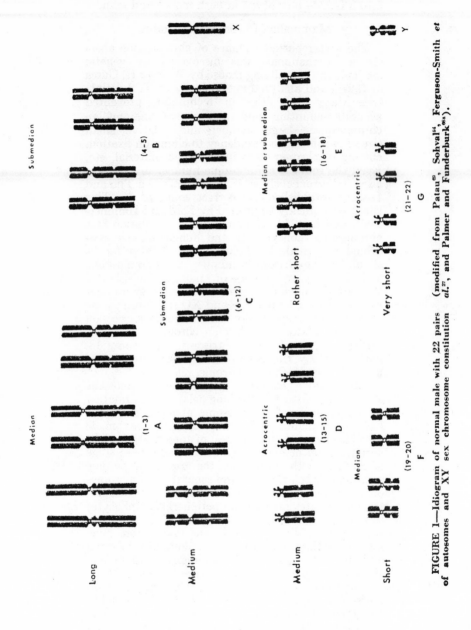

FIGURE 1—Idiogram of normal male with 22 pairs (modified from Patau[65], Sohval[84], Ferguson-Smith *et al.*[27], and Palmer and Funderburk[65a]). of autosomes and XY sex chromosome constitution

where differences may indicate abnormalities. An idiogram is a diagrammatic representation of the entire standard chromosome complement, showing their relative lengths, position of centromeres, arm ratios, satellites, secondary constrictions and other features. Figure 1 shows an idiogram of a normal human male with 22 pairs of autosomes and XY sex chromosome constitution. A karyotype is constructed from photographs of chromosomes which are arranged in pairs similar to the idiogram. Figure 2 shows a karyotype of a normal human female.

Incidence of Human Chromosome Anomalies

Chromosome anomalies are relatively frequent events. They have been estimated to occur in 0.48 percent of all newborn infants (one in 208)[81]. At least 25 percent of all spontaneous miscarriages result from gross chromosomal errors[13]. The general incidence of chromosome abnormalities in abortuses is more than fifty times the incidence at birth.

Although it is impossible to obtain an accurate total of victims suffering from effects of chromosome aberrations, one can make rough calculations on the basis of their estimated frequencies in the population of the United States assuming that there is no appreciable difference in life expectancy between these individuals and those with normal chromosome complements. Although this assumption probably is unjustifiable, it suffices for this rough calculation. Among the current population of 202 million we arrive at a figure of 1,136,971 total afflicted with chromosome abnormalities. This total probably represents an underestimate since it does not include all types of chromosome aberrations. Table I indicates totals for a number of specific syndromes.

Syndromes Related to Autosome Abnormalities

Down's syndrome

This defect results from duplication of all or part of autosome 21, either in the trisomic state or as a translocation to another chromosome, usually a 13–15 (D group) or 16–18 (E group) but may be to another G group chromosome. The overall incidence is about 1 in 700 live births[71], but the trisomic type is correlated with age of the mother, having a frequency of about 1 in 2000 in mothers under 30 years of age, and increasing to 1 in 40 in mothers aged 45 or over. The translocation type constitutes about 3.6 percent of cases and is unrelated to the

31

Table I. Total frequencies in the United States of various types of chromosomal abnormalities, calculated on the basis of 202 million current population and the estimated frequency of each abnormality. (It must be noted that the grand total does not include all types of chromosome aberrations, therefore must be lower than the real value)

Syndrome	Chromosome number	Estimated incidence	Calculated number in U.S.
Down's trisomy 21	47	1 in 700	288,571
Trisomy D	47	1 in 10,000	20,200
Trisomy E	47	1 in 4000	50,500
Trisomy X	47	1 in 10,000 females	101,000
Turner's XO	45	1 in 5000 females	20,200
Klinefelter's XXY	47	1 in 400 males	252,500
Double Y XYY	47	1 in 250 males	404,000
		Total	1,136,971

mother's age, but is transmitted in a predictable manner. Among mental retardates mongoloids represent 16.7 percent.

Clinical features include physical peculiarities ranging from slight anomalies to severe malformations in almost every tissue of the body. Typical appearance of a mongoloid shows slanting eyes, saddle nose, often a large ridged tongue that rolls over a protruding lip, a broad, short skull and thick, short hands, feet and trunk. Frequent complications occur: cataract or crossed eyes, congenital heart trouble, hernias, and a marked susceptibility to respiratory infections. They exhibit characteristic dermatoglyphic patterns on palms and soles. Also they have many biochemical deviations from normal, such as decreased blood-calcium levels and diminished excretion of tryptophane metabolites. Early ageing is common.

All mongoloids are mentally retarded; they usually are 3 to 7 years old mentally. Among the relatively intelligent patients, abstract reasoning is exceptionally retarded.

Female mongoloids are fertile and recorded pregnancies have yielded approximately 50 percent mongoloid offspring. Fortunately male mongoloids are sterile. Examination of their testes reveals varying degrees of spermatogenic arrest correlated with the abnormal chromosome features.

Among mongoloids there is a prevelence of leukemia in childhood; the incidence is some twenty times greater than in the general population. Simultaneous occurrence with other syndromes such as Klinefelter's, also is found, and many cases of mosaicism have been described.

E trisomy syndrome

This is another autosomal anomaly, which involves chromosomes 16, 17 and 18, and is estimated to occur at a frequency of 1 in 4000 live births[20]. Many others die before birth, thus contributing to the large number of miscarriages and stillbirths. These individuals survive only a short time, from one-half day to 1460 days, with an average of 239 days, but females live significantly longer than males.

Trisomy 17 syndrome

Many serious defects usually are present in afflicted invididuals[25]: odd shaped skulls, low-set and malformed ears, triangular mouth with receding chin, webbing of neck, shield-like chest, short stubby fingers, and toes with short nails, webbing of toes, ventricular septal defect and mental retardation, as well as abnormal facies, micrognathia and high arched palate.

Trisomy 18 syndrome

This anomoly[70, 82] is characterized by multiple congenital defects of which the most prominent clinical features are: mental retardation with moderate hypertonicity, low-set malformed ears, small

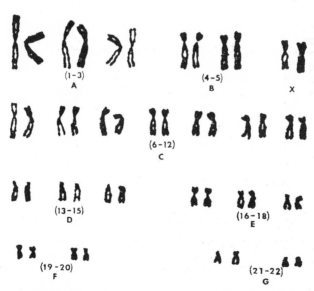

FIGURE 2—Karyotype of a normal human female with 22 pairs of autosomes and two X chromosomes.

mandible, flexion of fingers with the index finger overlying the third, and severe failure to thrive. It generally results in death in early infancy. Its frequency increases with advanced maternal age. Three times as many females as males have been observed; one would expect that more males with this syndrome will be found among stillbirths and fetal deaths.

D syndrome

This trisomy[19, 56, 70, 83] involves chromosomes 13, 14 and 15, and has an estimated frequency of about 1 in 10,000 live births. Many others die in utero. Survival time has been reported from 0 to 1000 days, with an average of 131 days.

Clinical features include: microcephaly, eye anomalies (corneal opacities, colobomata, microphthalmia, anophthalmia), cleft lip, cleft palate, brain anomalies (particularly arrhinencephaly), supernumerary digits, renal anomalies (especially cortical microcysts), and heart anomalies.

Trisomy 22 syndrome

This syndrome produces mentally retarded, schizoid individuals. Reports of its occurrence are too few to permit an estimate of its frequency in the population.

Cri-du-chat syndrome

Lejeune et al.[55] first described this anomaly in 1965, which involves a deficiency of the short arm of a B group chromosome, number 5. Translocations appear to be a common cause of the defect, an estimated 13 percent of cases being associated with translocations; described cases have had B/C, B/G, and B/D translocations[23]. The high proportion emphasizes the importance of unbalanced gamete formation in translocation heterozygotes as a cause of this syndrome. Among parents the frequency of male and female carriers is approximately the same, a situation that contrasts with the much greater frequency of female carriers of a D/G translocation among parents of translocation mongoloids.

Typical clinical features of cri-du-chat individuals are: low birth weight, severe mental retardation, microcephaly, hypertelorism, retrognathism, downward slanting eyes, epicanthal folds, divergent strabismus, growth retardation, narrow ear canals, pes planus and short metacarpals and metatarsals. About 25 to 30 percent of them have congenital heart disorders. A characteristic cat-like cry in infancy is responsible for the name of the syndrome. The cry is due to a small epiglottis and larynx and an atrophic vestibule. However, this major diagnostic sign disappears after infancy, making identification of older cases difficult.

An estimate of the frequency of this syndrome is given as over 1 percent but less than 10 percent of the severely mentally retarded patients. Many have IQ scores below 10, and most are institutionalized.

Philadelphia chromosome

Finally, among autosomal aberrations, a deleted chromosome 21 occurs in blood-forming stem cells in red bone marrow. This deletion, which shows up long after birth, appears to be the primary event causing chronic granulocytic leukemia. This aberration was discussed in 1960 by Nowell and Hungerford[67] (also Baikie et al.[3]).

Syndromes Related to Sex Chromosome Aberrations

The great majority of known chromosomal abnormalities in man involve the sex chromosomes. In one survey (that excluded XYY) it was estimated that abnormalities occurred in 1 out of every 450 births; if the recent estimate of XYY[81] is correct, the frequency actually is much higher. Increased knowledge about sex chromosome aberrations is probably related to the greater concentration of attention on patients with sexual disorders, but is due in part to the ability to detect carriers of an extra X chromosome by the so-called sex chromatin body or Barr body[6]. This structure is a stainable granule at the periphery of a resting nucleus and, according to the Lyon hypothesis[57], is considered to be an inactivated X chromosome. A normal female cell has one Barr body, since it has two X chromosomes, and is said to be sex chromatin positive (or one positive). A normal male cell has no Barr body and is said to be sex chromatin negative.

Klinefelter's syndrome

The first sex chromosome anomaly described in 1959 by Jacobs and Strong[46] and also by Ford et al.[29] was the XXY constitution that is typical of Klinefelter's syndrome. Buccal smears from these patients are sex chromatin positive. They can be tentatively diagnosed by this test along with clinical symptoms. Final confirmation of diagnosis can be achieved by karyotype analysis using either bone marrow aspiration or peripheral blood culture.

Victims of Klinefelter's syndrome are always male

but they are generally underdeveloped, eunachoid in build, with small external genitalia, very small testes and prostate glands, with underdevelopment of hair on the body, pubic hair and facial hair, frequently with enlarged breasts (gynecomastia), and many have a low IQ.

The classical type with two X chromosomes and one Y chromosome was the first case discovered, but subsequently chromosome compositions of XXXY, XXXXY, XXYY[66] and XXXYY[7, 8, 63, 77] have been reported. In addition, numerous mosaics have been described, including double, triple and quadruple numeric mosaics, as well as combinations of numeric and structural mosaics. These conditions are summarized in Table II. They all resemble the XXY Klinefelter's phenotypically and are considered modified Klinefelter's syndromes. The classical XXY type may have low normal mental development or may be retarded, but other types show increasingly greater mental retardation.

The incidence of Klinefelter's syndrome is estimated to be 1 in 400 male live births, which represents from 1 to 3 percent of mentally deficient patients. This condition also has been correlated with age of the mother: the older the mother, the greater the risk of having such a child. These individuals usually are sterile. Spermatogenesis is generally totally absent. Hyalinization of the semeniferous tubules begins shortly before puberty. Congenital malformations are rare. Mental retardation is present in approximately 25 percent of affected individuals, and mental illness may be more common than in the general population.

Turner's syndrome

Female gonadal dysgenesis was described by Turner in 1938 as a syndrome of primary amenorrhea, webbing of the neck, cubitas valgus and short stature, coarctation of aorta, failure of ovarian development and hormonal abnormalities. Patients exhibit sexual infantilism; their breasts are usually underdeveloped, nipples often widely spaced, particularly in those subjects who have a shield or funnel chest deformity. Usually sexual hair is scanty; external genitalia are infantile; labia small or unapparent; clitoris usually normal, although may be enlarged. The uterus is infantile; the tubes long and narrow; the gonads represented by long, narrow, white streaks of connective tissue in normal position of ovary. They are almost always sterile. Hormonal secretions usually are abnormal. Short-

Table II. Reported sex chromosomal constitutions in Klinefelter's syndrome (modified from Reitalu[77])

		Sex chromosomal constitution			
Only one karyotype observed per individual		XXY			
		XXYY			
		XXXY			
		XXXYY			
		XXXXY			
Numeric mosaics	Double	XX	XXY		
		XY	XXY		
		XY	XXXY		
		XXY	XXYY		
		XXXY	XXXXY		
		XXXX	XXXXY		
	Triple	XY	XXY	XXYY	
		XX	XXY	XXXY	
		XY	XXY	XXXY	
		XO	XY	XXY	
		XX	XY	XXY	
		XXXY	XXXXY	XXXXYY	
		XXXY	XXXXY	XXXXXY	
	Quadruple	XXY	XY	XX	XO
Numeric and structural mosaics	Double	XXY	XXxY		
	Triple	XY	XXY	XXxY	
		XxY	Xx	XY	

ness of stature is characteristic and many other skeletal abnormalities occur. Peculiar facies include small mandible, anti-mongolian slant of eyes, depressed corners of mouth, low-set ears, auricles sometimes deformed. Cardiovascular defects are frequent, the most common being coarctation of the aorta. Slight intellectual impairment is found in some patients, particularly those with webbing of the neck.

In 1954 it was discovered that many patients with ovarian agenesis were sex chromatin negative, and in 1959 Ford and colleagues[30] gave the first chromosome analysis showing that Turner's syndrome has the sex chromosome abnormality of only one X chromosome (XO) rather than two X's. It was quickly confirmed by Jacobs and Keay[45] and by Fraccaro et al[33].

Mosaicism is known to exist—both 45 chromosome cells and 46 chromosome cells occur side by side in tissues of the individual—and can result from nondisjunction in early embryonic development. Isochromosomes sometimes are involved, e.g., creating a situation with 3 long arms of the X chromosome but only 1 short arm.

The incidence of XO Turner's syndrome is estimated as 1 in approximately 5000 women; many die in utero.

Large scale screening of newborn babies by buccal smears can permit detection of chromatin negative females, chromatin positive males, .and double, triple, quadruple and quintuple positive cases of either sex[58]. Table III shows the relationship between sex chromosome complements and sex chromatin pattern.

Triplo-X syndrome

Females containing three[47], four and five X chromosomes are known[1, 47, 63]. The triplo-X syndrome is thought to have an incidence of about 1 in 800 live female births. This syndrome was first described by Jacobs et al.[47] in 1959. Although it has no distinctive clinical picture, menstrual irregularlties may be present, secondary amenorrhea or premature menopause. Most cases have no sexual abnormalities and many are known to have children. The most characteristic feature of 3X females is mental retardation. Qudruple-[14] and quintuple-X[50] syndromes are much rarer. These individuals are mentally retarded, usually the more X chromosomes present, the more severe the retardation. Frequently these individuals are fertile.

An extra X chromosome confers twice the usual risk of being admitted to a hospital with some form of mental illness. The loss of an X, on the other hand, has no association with mental illness; thus the chance of mental hospital admission is not raised for an XO female. An extra X chromosome also predisposes to mental subnormality. The prevalence of psychosis among patients in hospitals for the subnormal is unusually high in males with two or more X chromosomes.

Numerous other sex chromosome anomalies occur[38], many involving mosaics and structural chromosome aberrations. For example, occasionally an XY embryo will differentiate into a female, a situation referred to as testicular feminization male pseudohermaphrodite (Morris syndrome)[60]. These individuals have only streak gonads and vestigial internal genital organs. They usually have undeveloped breasts and do not menstruate. They are invariably sterile[76].

Still other sexual abnormalities are intersexes and true hermaphrodites, many of which have an XX sex chromosome constitution or are mosaics for sex chromosomes such as XO/XY or XX/XY or more complicated mixtures[31]. Sex chromosome mosaicism is very common. Almost every sex chromosome combination found alone has been found in association with one or more cell lines with a different sex chromosome constitution. These mosaics exhibit quite a variable expression; for example in an XO/XY mosaic the external genitalia can appear female, male or intersexual[85].

The YY syndrome

The male with an extra Y chromosome (XYY) has attracted much attention in the public press as well as in scientific circles because of his reputed antisocial, aggressive and criminal tendencies[1, 2, 64]. Although this abnormality belongs in the above category of syndromes related to sex chromosome aberrations, it has been singled out for special discussion because of its social and legal implications.

Evidence supporting the existence of a double Y syndrome has accumulated within the last six years. Studies in Sweden[32] showed an unusually large number of XXYY and XYY men among hard-to-manage patients in mental hospitals. These observations received impressive confirmation in studies of maximum security prisons and hospitals for the criminally insane in Scotland where an astonishingly high frequency (2.9 percent) of XYY males were found[48]. This was over fifty times higher than the then current estimate of 1 in 2000 in the general population. Subsequently many additional studies on the YY syndrome have appeared and a composite picture of the XYY male emerged[5, 19, 15, 21a, 34, 35, 41, 72-75, 78, 80, 92].

The principal features of the extra Y syndrome appear to be exceptional height and a serious personality disorder leading to behavioral disturbances. It seems likely it is the behavior disorder rather than their intellectual incompetence that prevents them from functioning adequately in society[18].

Clinically the XYY males are invariably tall (usually six feet or over) and frequently of below-average intelligence. They are likely to have unusual sexual tastes, often including homosexuality. A history of antisocial behavior, violence and conflict with the police and educational authorities from early years is characteristic[86] of the syndrome.

Although these males usually do not exhibit obvious physical abnormalities[12, 24, 40, 42, 52, 91], several cases of hypogonadism[11], some with undescended testes, have been reported. Others have epilepsy, malocclusion and arrested development[87], but these symptoms may be fortuitously associated. One case was associated with trisomy 21[61], another with pseudohermaphrodism[35]. The common feature of an acne-scarred face may be related to altered hormone production. The criminally aggressive group were found to have evidence of an increased androgenic steroid production as reflected by high plasma and urinary testosterone levels[12, 43]. If the high level of plasma testosterone is characteristic of XYY individuals, it suggests a mechanism through which this condition may produce behavioral changes, possibly arising at puberty.

Antisocial and aggressive behavior in XYY individuals may appear early in life, however, as evidenced by a case reported by Cowie and Kahn[22]. A prepubertal boy with normal intelligence, at the age of 4½ years, was unmanageable, destructive, mischievous and defiant, overadventurous and without fear. His moods alternated; there were sudden periods of overactivity at irregular intervals when he would pursue his particular antisocial activity with grim intent. Between episodes he appeared happy and constructive. The boy was over the 97th percentile in height for his age, a fact that supports the view that increased height in the XYY syndrome is apparent before puberty.

It has been suggested that the ordinary degree of aggressiveness of a normal XY male is derived from his Y chromosome, and that by adding another

Table III. The relationship between sex, sex chromosome complement and sex chromatin pattern (modified from Miller[63])

Sex chromatin pattern	Sexual phenotype	
	Female	Male
−	XO	XY
−	XY (testicular feminization)	XYY
+	XX	XXY
		XXYY
++	XXX	XXXY
		XXXYY
+++	XXXX	XXXXY
++++	XXYYX	

Y a double dose of those potencies may facilitate the development of aggressive behavior[65] under certain conditions. A triple dose (XYYY) would be present in the case reported by Townes et al.[90].

40

The first reported case of an XYY constitution[39, 79] was studied because the patient had several abnormal children, although he appeared to be normal himself. Until recently, reports of the XYY constitution have been uncommon, probably because no simple method exists for screening the double-Y condition that is comparable to the buccal smear—sex chromatin body technique for detecting an extra X chromosome. Another possible explanation for the rarity of reports on the XYY karyotype is the absence of a specific phenotype in connection with it. Most syndromes with a chromosome abnormality are ascertained because of some symptom or clinical sign that indicates a need for chromosome analysis. Consequently there have been few studies that place the incidence of this chromosome abnormality in its proper perspective to the population as a whole.

Very recently a study of the karyotypes of 2159 infants born in one year was made by Sergovich et al.[31]. These investigators detected 0.48 percent of gross chromosome abnormalities. In this sample the XYY condition appeared in the order of 1 in 250 males, which would make it the most common form of aneuploidy known for man. The previous estimate was about 1 in 2000 males. If this figure of 1/250 is valid for the population as a whole, it means that the great majority of cases go undetected and consequently must be phenotypically normal and behave near enough to the norm to go unrecognized.

Several cases of asymptomatic males have been published, including the first one described (Sandberg et al.[79] and Hauschka et al.[39]), which proved to be fertile. It appears that the sons of XYY men do not inherit their father's extra Y chromosome[59a].

Another fertile XYY male, reported by Leff and Scott[53], had inferiority feelings, was slightly hypochondriacal and obsessional, and not very aggressive. He gave a general impression of emotional immaturity. He was 6 feet, 6 inches tall, healthy, with normal genitalia and electroencephalogram. His IQ was 118. Wiener and Sutherland[93] discovered by chance an XYY male who was normal; he was 5 feet, 9½ inches tall, with normal genitalia and body hair, normal brain waves, and with an IQ of 97. He exhibited a cheerful disposition and mild temperment, had no apparent behavioral disturbance and never required psychiatric advice. This case supports the idea that an XYY male can lead a normal life.

Social and Legal Implications of the YY Syndrome

The concept that when a human male receives an

extra Y chromosome it may have an important and potentially antisocial effect upon his behavior is supported by impressive evidence[15, 21]. Lejuene states that "There are no born criminals but persons with the XYY defect have considerably higher chances." Price and Whatmore[74] describe these males as psychopaths, "unstable and immature, without feeling or remorse, unable to construct adequate personal relationships, showing a tendency to abscond from institutions and committing apparently motiveless crimes, mostly against property." Casey and coworkers[16] examined the chromosome complements in males 6 feet and over in height and found: 12 XYY among 50 mentally subnormal and 4 XYY among mentally ill patients detained because of antisocial behavior; also 2 XYY among 24 criminals of normal intelligence. They concluded that their results indicate that an extra Y chromosome plays a part in antisocial behavior even in the absence of mental subnormality. The idea that criminals are degenerates because of bad heredity has had wide appeal. There is no doubt that genes do influence to some extent the development of behavior. The influence may be strongly manifested in some cases but not in others. Some individuals appear to be driven to aggressive behavior.

Several spectacular crime cases served to publicize this genetic syndrome, and it has been played up in newspapers, news magazines, radio and television. In 1965 Daniel Hugon, a stablehand, was charged with the murder of a prostitute in a cheap Paris hotel. Following his attempted suicide he was found to have an XYY sex chromosome constitution. Hugon surrendered to the police and his lawyers contended that he was unfit to stand trial because of his genetic abnormality. The prosecution asked for five to ten years; the jury decided to give him seven.

Richard Speck, the convicted murderer of eight nurses in Chicago in 1966, was found to have an XYY sex chromosome constitution. He has all the characteristics of this syndrome found in the Scottish survey: he is 6 feet 2 inches tall, mentally dull, being semiliterate with an IQ of 85, the equivalent of a 13-year-old boy. Speck's face is deeply pitted with acne scars. He has a history of violent acts against women. His aggressive behavior is attested by his record of over 40 arrests. Speck was sentenced to death but the execution has been held up pending an appeal of the conviction.

In Melbourne, Australia, Lawrence Edward Hannell, a 21-year-old laborer on trial for the stabbing of a 77-year-old widow, faced a maximum

42

sentence of death. He was found to have an XYY constitution, mental retardation, an aberrant brain wave pattern, and a neurological disorder. Hannel pleaded not guilty by reason of insanity, and a criminal court jury found him not guilty on the ground that he was insane at the time of the crime.

A second Melbourne criminal with an XYY constitution, Robert Peter Tait, bludgeoned to death an 81-year-old woman in a vicarage where he had gone seeking a handout. He was convicted of murder and sentenced to hang, but his sentence was commuted to life imprisonment.

Another case is that of Raymond Tanner, a convicted sex offender, who pleaded guilty to the beating and rape of a woman in California. He is 6 feet 3 inches tall, mentally disordered, and has an XYY complement. A superior court judge is attempting to decide whether Tanner's plea of guilty to assault with intent to commit rape will stand, or whether he will be allowed to plead innocent by reason of insanity.

Criminal lawyers in the United States have already begun to request genetic studies of their clients. In October of 1968 a lawyer for Sean Farley, a 26-year-old XYY man in New York who was charged with a rape-slaying, maneuvered to raise the issue of his client's genetic defect in court.

Many questions are raised by the double Y syndrome—basic social, legal and ethical questions—which will become more and more insistent as the implications of chromosome abnormalities take root in the public mind. Is an extra Y chromosome causally related to antisocial behavior? Is there a genetic basis for criminal behavior? If a man has an inborn tendency toward criminal behavior, can we fairly hold him legally accountable for his acts? If a criminal's chromosomes are at fault, how can we rehabilitate him?

The evidence to date is inadequate to prove conclusively the validity of the syndrome and convict all of the world's estimated five million XYY males of innate aggressive or criminal tendencies. But if the concept is proved, what then? The first step would seem to be to identify the XYY infants in the general population. This suggests the need for a nationwide program of automatic chromosome analysis of all newborns.

How should society deal with XYY individuals? If they are genetically abnormal, they should not be treated as normal. If the XYY condition dooms a man to a life of crime, he should be restrained but not punished. Mongolism also is a chromosome abnormality, and afflicted individuals are not held responsible for their behavior. Some valuable suggestions on the legal aspects of the double Y syndrome have been published recently by Kennedy McWhirter[59]. Elsewhere, Kessler and Moos[51] claim that definitive concepts relating to the YY syndrome have been accepted prematurely.

If all infants could be karyotyped at birth or soon after, society could be forearmed with information on chromosome abnormalities and perhaps it could institute the proper preventive and other measures at an early age. Although society can not control the chromosomes (at least at the present time) it can do a great deal to change certain environmental conditions that may encourage XYY individuals to commit criminal acts.

The theory that a genetic abnormality may predispose a man to antisocial behavior, including crimes of violence, is deceptively and attractively simple, but will be difficult to prove. Extensive chromosome screening with prospective follow-up of XYY males will be essential to determine the precise behavioral risk of this group. It is by no means universally accepted yet. Many geneticists urge that we should be cautious in accepting the interpretation that the double Y condition is specifically associated with criminal behavior, and particularly so with reference to the medicolegal validity of these concepts.

Literature Cited

1. ANONYMOUS. The YY syndrome. *Lancet* 1:583–584. 1966.
2. ————. Criminal behavior—XYY criterion doubtful. *Science News* 96:2. 1969.
3. BAIKIE, A. G., W. M. COURT BROWN, K. E. BUCKTON, D. G. HARNDEN, P. A. JACOBS, and I. M. TOUGH. A possible specific chromosome abnormality in human chronic myeloid leukemia. *Nature* 188:1165–1166. 1960.
4. ————, O. MARGARET GARSON, SANDRA M. WESTE, and JEAN FERGUSON. Numerical abnormalities of the X chromosome. *Lancet* 1:398–400. 1966.
5. BALODIMOS, MARIOS C., HERMANN LISCO, IRENE IRWIN, WILMA MERRILL, and JOSEPH F. DINGMAN. XYY karyotype in a case of familial hypogonadism. *J. Clin. Endocr.* 26:443–452. 1966.
6. BARR, M. L. Sex chromatin and phenotype in man. *Science* 130:679. 1959.
7. ———— and D. H. CARR. Sex chromatin, sex chromosomes and sex anomalies. *Canad. Med. Assn. J.* 83:979–986. 1960.
8. ————, D. H. CARR, H. C. SOLTAN, RUTH G. WIENS, and E. R. PLUNKETT. The XXYY variant of Klinefelter's syndrome. *Canad. Med. Assn. J.* 90:575–580. 1964.
9. BELSKY, JOSEPH L. and GEORGE H. MICKEY. Human cytogenetic studies. *Danbury Hospital Bull.* 1:19–20. 1965.

10. BOCZKOWSKI, K. and M. D. CASEY. Pattern of DNA replication of the sex chromosomes in three males, two with XYY and one with XXYY karyotype. *Nature* 213:928–930. 1967.

11. BUCKTON, KARIN.E., JANE A. BOND, and J. A. McBRIDE. An XYY sex chromosome complement in a male with hypogonadism. *Human Chromosome Newsletter* No. 8, p. 11. Dec. 1962.

12. CARAKUSHANSKY, GERSON, RICHARD L. NEU, and LYTT I. GARDNER. XYY with abnormal genitalia. *Lancet* 2:1144. 1968.

13. CARR, D. H. Chromosome studies in abortuses and stillborn infants. *Lancet* 2:603–606. 1963.

14. ————, M. L. BARR, and E. R. PLUNKETT. An XXXX sex chromosome complex in two mentally defective females. *Canad. Med. Assn. J.* 84:131–137. 1961.

15. CASEY, M. D., C. E. BLANK, D. R. K. STREET, L. J. SEGALL, J. H. McDOUGALL, P. J. McGRATH, and J. L. SKINNER. YY chromosomes and antisocial behavior. *Lancet* 2:859–860. 1966.

16. ————, L. J. SEGALL, D. R. K. STREET, and C. E. BLANK. Sex chromosome abnormalities in two state hospitals for patients requiring special security. *Nature* 209:641–642. 1966.

17. ————, D. R. K. STREET, L. J. SEGALL, and C. E. BLANK. Patients with sex chromatin abnormality in two state hospitals. *Ann. Human Genet.* 32:53–63. 1968.

18. CLOSE, H. G., A. S. R. GOONETILLEKE, PATRICIA A. JACOBS, and W. H. PRICE. The incidence of sex chromosomal abnormalities in mentally subnormal males. *Cytogenetics* 7:277–285. 1968.

19. CONAN, P. E. and BAYZAR ERKMAN. Frequency and occurrence of chromosomal syndromes. I. D-trisomy. *Am. J. Human Genet.* 18:374–386. 1966.

20. ———— and ————. Frequency and occurrence of chromosomal syndromes. II. E-trisomy. *Am. J. Human Genet.* 18:387–398. 1966.

21. COURT BROWN, W. M. Sex chromosomes and the law. *Lancet* 2:508–509. 1962.

21a. ————. Males With an XYY sex chromosome complement. *J. Med. Genet.* 5:341–359. 1968.

22. COWIE, JOHN and JACOB KAHN. XYY constitution in prepubertal child. *Brit. Med. J.* 1:748–749. 1968.

23. DECAPOA, A., D. WARBURTON, W. R. BREG, D. A. MILLER, and O. J. MILLER. Translocation heterozygosis: a cause of five cases of *cri du chat* syndrome and two cases with a duplication of chromosome number five in three families. *Am. J. Human Genet.* 19:586–603. 1967.

24. DENT, T., J. H. EDWARDS, and J. D. A. DELHANTY. A partial mongol. *Lancet* 2:484–487. 1963.

25. EDWARDS, J. H., D. G. HARNDEN, A. H. CAMERON, V. MARY CROSSE, and O. H. WOLFF. A new trisomic syndrome. *Lancet* 1:787–790. 1960.

26. EGGEN, ROBERT R. Chromosome Diagnostics in Clinical Medicine. Charles C. Thomas, Springfield, Ill. 1965.

27. FERGUSON-SMITH, M. A., MARIE E. FERGUSON-SMITH, PATRICIA M. ELLIS, and MARION DICKSON. The sites and relative frequencies of secondary constrictions in human somatic chromosomes. *Cytogenetics* 1:325–343. 1962.

28. FORD, C. E. and J. L. HAMERTON. The chromosomes of man. *Nature* 178:1020–1023. 1956.

29. ————, K. W. JONES, O. J. MILLER, URSULA MITTWOCH, L. S. PENROSE, M. RIDLER, and A. SHAPIRO. The chromosomes in a patient showing both mongolism and the Klinefelter syndrome. *Lancet* 1:709–710. 1959.

45

30. ————, K. W. Jones, P. E. Polani, J. C. deAlmedia, and J. H. Briggs. A sex-chromosome anomaly in a case of gonadal dysgenesis (Turner's syndrome). *Lancet* 1:711–713. 1959.

31. ————, P. E. 'Polani, J. H. Briggs, and P. M. F. Bishop. A presumptive human XXY/XX mosaic. *Nature* 183:1030–1032. 1959.

32. Forssman, H. and G. Hambert. Incidence of Klinefelter's syndrome among mental patients. *Lancet* 1:1327. 1963.

33. Fraccaro, M., K. Kaijser, and J. Lindsten. Chromosome complement in gonadal dysgenesis (Turner's syndrome). *Lancet* 1:886. 1959.

34. ————, M. Glen Bott, P. Davies, and W. Schutt. Mental deficiency and undescended testia in two males with XYY sex chromosomes. *Folia Hered. Pathol.* (Milan) 11:211–220. 1962.

35. Franks, Robert C., Kenneth W. Bunting, and Eric Engel. Male pseudohermaphrodism with XYY sex chromosomes. *J. Clin. Endocr.* 27:1623–1627. 1967.

36. Fraser, J. H., J. Campbell, R. C. MacGillivray, E. Boyd, and B. Lennox. The XXX syndrome—frequency among mental defectives and fertility. *Lancet* 2:626–627. 1960.

37. Gates, William H. A case of non-disjunction in the mouse. *Genetics* 12:295–306. 1927.

38. Hamerton, J. L. Sex chromatin and human chromosomes. *Intern. Rev. Cytol.* 12:1–68. 1961.

39. Hauschka, Theodore S., John E. Hasson, Milton N. Goldstein, George F. Koepf, and Avery A. Sandberg. An XYY man with progeny indicating familial tendency to non-disjunction. *Am. J. Human Genet.* 14:22–30. 1962.

40. Hayward, M. D. and B. D. Bower. Chromosomal trisomy associated with the Sturge-Weber syndrome. *Lancet* 2:844–846. 1960.

41. Hunter, H. Chromatin-positive and XYY boys in approved schools. *Lancet* 1:816. 1968.

42. Hustinx, T. W. J. and A. H. F. van Olphen. An XYY chromosome pattern in a boy with Marfan's syndrome. *Genetica* 34:262. 1963.

43. Ismail, A. A. A., R. A. Harkness, K. E. Kirkham, J. A. Loraine, P. B. Whatmore, and R. P. Brittain. Effect of abnormal sex-chromosome complements on urinary testosterone levels. *Lancet* 1:220–222. 1968.

44. Jacobs, P. A., A. G. Baikie, W. M. Court Brown, and J. A. Strong. The somatic chromosomes in mongolism. *Lancet* 1:710. 1959.

45. ———— and A. J. Keay. Chromosomes in a child with Bonnevie-Ullrich syndrome. *Lancet* 2:732. 1959.

46. ———— and J. A. Strong. A case of human intersexuality having a possible XXY sex-determining mechanism. *Nature* 182:302–303. 1959.

47. ————, A. G. Baikie, W. M. Court Brown, T. N. MacGregor, N. MacLean, and D. G. Harnden. Evidence for the existence of the human "super female". *Lancet* 2:423–425. 1959.

48. ————, Muriel Brunton, Marie M. Melville, R. P. Brittain, and W. F. McClemont. Aggressive behaviour, mental subnormality and the XYY male. *Nature* 208: 1351–1352. 1965.

49. ————, W. H. Price, W. M. Court Brown, R. P. Brittain, and P. B. Whatmore. Chromosome studies on men in a maximum security hospital. *Ann. Human Genet.* 31:330–347. 1968.

50. Kesaree, Nirmala and Paul V. Woolley. A phenotypic female with 49 chromosomes, presumably XXXXX. *J. Pediat.* 63:1099–1103. 1963.

51. KESSLER, SEYMOUR and RUDOLPH H. Moos. XYY chromosome: premature conclusions. *Science* 165:442. 1969.

52. KOSENOW, W. and R. A. PFEIFFER. YY syndrome with multiple malformations. *Lancet* 1:1375-1376. 1966.

53. LEFF, J. P. and P. D. SCOTT. XYY and intelligence. *Lancet* 1:645. 1968.

54. LEJEUNE, J., M. GAUTIER, and R. TURPIN. Etude des chromosomes somatiques de neuf enfants mongoliens. *Compt. Rend. Acad. Sci.* 248:1721-1722. 1959.

55. ————, J. LAFOURCADE, R. BERGER and M. O. RETHORE. Maladie du cri du chat et sa reciproque. *Ann. Genet.* 8:11-15. 1965.

56. LUBS, H. A., JR., E. V. KOENIG, and L. H. BRANDT. Trisomy 13-15: A clinical syndrome. *Lancet* 2:1001-1002. 1961.

57. LYON, M. F. Gene action in the X-chromosome of the Mouse (*Mus musculus* L.). *Nature* 190:372-373. 1961.

58. MACLEAN, N., D. G. HARNDEN, W. M. COURT BROWN, JANE BOND, and D. J. MANTLE. Sex-chromosome abnormalities in newborn babies. *Lancet* 1:286-290. 1964.

59. MCWHIRTER, KENNEDY. XYY chromosome and criminal acts. *Science* 164:1117. 1969.

59a. MELNYK, JOHN, FRANK VANASEK, HAVELOCK THOMPSON, and ALFRED J. RUCCI. Failure of transmission of supernumerary Y chromosomes in man. Abst. Am. Soc. Human Genet. Annual Meeting, Oct. 1-4, 1969.

60. MICKEY, GEORGE H. Chromosome studies in testicular feminization syndrome in human male pseudohermaphrodites. *Mammalian Chromosome Newsletter* No. 9, p. 60. 1963.

61. MIGEON, BARBARA R. G trisomy in an XYY male. *Human Chromosome Newsletter* No. 17. Dec. 1965.

62. MILCU, M., I. NIGOESCU, C. MAXIMILIAN, M. GAROIU, M. AUGUSTIN, and ILEANA ILIESCU. Baiat cu hipospadias si cariotip XYY. *Studio si Cercetari de Endocrinologie* (Bucharest) 15:347-349. 1964.

63. MILLER, ORLANDO J. The sex chromosome anomalies. *Am. J. Obstet. Gynec.* 90:1078-1139. 1964.

64. MINCKLER, LEON S. Chromosomes of criminals. *Science* 163:1145. 1969.

65. MONTAGU, ASHLEY. Chromosomes and crime. *Psychology Today* 2:43-49. 1968.

66. MULDAL, S. and C. H. OCKEY. The "double male": a new chromosome constitution in Klinefelter's syndrome. *Lancet* 2:492-493. 1960.

67. NOWELL, P. C. and D. A. HUNGERFORD. A minute chromosome in human granulocytic leukemia. *Science* 132:1497. 1960.

68. PAINTER, T. S. The chromosome constitution of Gates "non-disjunction" (v-o) mice. *Genetics* 12:379-392. 1927.

68a. PALMER, CATHERINE G. and SANDRA FUNDERBURK. Secondary constrictions in human chromosomes. *Cytogenetics* 4:261-276. 1965.

69. PATAU, K. The identification of individual chromosomes, especially in man. *Am. J. Human Genet.* 12:250-276. 1960.

70. ————, D. W. SMITH, E. THERMAN, S. L. INHORN, and H. P. WAGNER. Multiple congenital anomalies caused by an extra chromosome. *Lancet* 1:790-793. 1960.

71. PENROSE, L. S. The Biology of Mental Defect. Grune and Stratton, New York. 1949.

72. PERGAMENT, EUGENE, HIDEO SATO, STANLEY BERLOW, and RICHARD MINTZER. YY syndrome in an American negro. *Lancet* 2:281. 1968.

73. PFEIFFER, R. A. Der Phanotyp der Chromosomenaberration XYY. *Wochenschrift* 91:1355-1256. 1966.

47

74. PRICE, W. H. and P. B. WHATMORE. Behaviour disorders and the pattern of crime among XYY males identified at a maximum security hospital. *Brit. Med. J.* 1:533. 1967.

75. ———, J. A. STRONG, P. B. WHATMORE, and W. F. McCLEMENT. Criminal patients with XYY sex-chromosome complement. *Lancet* 1:565–566. 1966.

76. PUCK, T. T., A. ROBINSON, and J. H. TJIO. A familial primary amenorrhea due to testicular feminization. A human gene affecting sex differentiation. *Proc. Exper. Biol. Med.* 103:192–196. 1960.

77. REITALU, JUHAN. Chromosome studies in connection with sex chromosomal deviations in man. *Hereditas* 59:1–48. 1968.

78. RICCI, N. and P. MALACARNE. An XYY human male. *Lancet* 1:721. 1964.

GETTING DOWN TO BUSINESS IN GENETICS:
RESEARCH, DIAGNOSIS, TREATMENT

Paul T. Libassi

"There are good reasons why there has been such a jump
in molecular biology as applied to bacterial viruses and
bacteria. One of these reasons, and possibly the most important, is the fact that these are problems on which work
can be done, by and large, by one man in a small laboratory with limited equipment. The nature of the universities and the nature of most research institutions are of
such quality that they insist a man work precisely in this
way. Therefore, large problems, large problems that involve coordinated efforts, very often are not pursued because the facilities and funding and cooperation are not
organized within the universities and research institutions; the result is that certain kinds of large problems
are selected against."

Thus, Dr. Seymour Cohen of the University of Pennsyl-
vania summarized the cardinal reason behind genetic science's failure to advance substantially beyond the plateau
of bacteria and bacteriophages. Financial requisites and
organizational requirements must be established, he emphasized at a National Institute of General Medical Sciences
symposium over two years ago, if the thrust of genetics
research is to progress beyond the realm of lower organisms toward more pertinent vistas.

That inherited diseases of man are indeed such a
"large problem" is attested to by medical statistician
Gabriel Stickle, who has calculated the numbers of life
years lost due to birth defects -- both inborn errors of
metabolism and structural deformities. Stickle has projected over 36 million life years lost in 1967 alone. According to that statistic, birth defects claim approximately 4.5 times as many life years as heart disease, 8
times as many as cancer, and about 10 times as many as
stroke.

Facilities and cooperation, financial requisites and
organizational requirements, remarked Dr. Cohen, are required to move the course of genetics research forward.
Through grants totaling in excess of $3.3 million, NIGMS
this past June established and funded a network of seven
Genetics Centers in five states (see table). Interdisciplinary analysis, cooperative endeavors, aggregates of
highly coordinated and integrated research projects characterize the strategem of this research conglomerate. The
design of each center prescribes that teams of scientists

LABORATORY MANAGEMENT, 1972, Vol. 10, No. 8, pp. 24-30, 55.

NIH GENETICS CENTERS PROGRAM

CENTER	PRINCIPAL INVESTIGATOR	DISEASE FOCI
Albert Einstein College of Medicine, Yeshiva University, New York ($319,300)	Harold M. Nitowsky, M.D.	Blood disorders; hereditary factors in diabetes
The Johns Hopkins University, Baltimore ($570,455)	Victor McKusick, M.D.	Bone and skeltal disorders, Marfan syndrome (heart enlargement, aortic weakness)
Mount Sinai School of Medicine, New York ($331,425)	Kurt Hirschhorn, M.D.	Hereditary kidney diseases, genetic eye diseases in children
University of California School of Medicine, San Diego ($719,659)	J. Edwin Seegmiller, M.D.	Metabolic disorders (Leson-Nynan, Tay-Sachs diseases); polysaccharide disorders (Hurler's syndrome)
University of California Medical Center at San Francisco ($480,793)	Charles J. Epstein, M.D.	Sex chromosome defects (Turner's syndrome); gout (disease of purine metabolism)
University of Texas Graduate School of Biomedical Sciences, Houston ($583,047)	Margery W. Shaw, M.D.	Carbohydrate storage disease; cystic fibrosis (identification of carriers)
University of Washington School of Medicine, Seattle ($958,207)	Arno G. Motulsky, M.D.	Sickle cell anemia (cyanate therapy); hereditary lipid patterns in heart diseases

Table courtesy of the National Institutes of Health

and medical geneticists explore the basic molecular nature of genetic diseases concurrent with clinical evaluations and population analyses involving affected patients and members of their families.

Studies at these Centers will be augmented by research findings generated by some 300 smaller projects which NIGMS also supports, bringing its total program expenditure to approximately $30 million per year. From these smaller projects -- the bulk of which deal with genetic investigation of lower organisms -- come the concepts and prototypes for the exploration of human genetics.

CENTERS VS GRANTS

Why does a funding instrument make the decision to found a Center requiring hundreds of thousands of dollars to support, when a smaller project can be sustained through an endowment of some $10,000 or $20,000? In a series of interviews with Dr. Fred H. Bergmann, Chief of the Genetics Section, Research Grants Branch at NIGMS, the answers to that and other questions regarding the evolution and rationale of these Centers were discussed.

The reasons why these Centers were established can be summarized under two general headings: managerial and scientific.

"The scientific potentialities and the talent to exploit them," explains Dr. Bergmann, "existed before NIGMS decided to fund these Centers. The problem was to catalyze these talents to interact as a group, to convince each separate discipline to function together. NIGMS is the glue that cements these parts into a whole."

But problems of management extend well beyond the absence of interdisciplinary communications and interaction. Difficulties that stem from the nature of individualized research -- Dr. Cohen calls them sociological problems in the discovery-making process -- must be contended with. The collaborative design of these seven Genetics Centers may well have done just that.

When dealing, for instance, with lower organisms such as *E. Coli.* which the majority of genetics research to date has, work can be done and great progress made by "one man in a small laboratory with limited equipment." The structural intricacies and complexities of human genetics, however, defy research of this type. They demand the professional acumen of an assemblage of specialists from the molecular biologist to the biochemist, enhanced by a sophisticated battery of equipment and facilities -- equipment and facilities which can easily exceed economic feasibility under a $10,000 stipend.

"You simply cannot justify expensive computer time or a tissue culture laboratory when you are dealing with a $10,000 grant," says Dr. Bergmann. To justify assests and facilities that run into large sums of money and the finances to implement them, he continues, interdisciplinary efforts concerned with broader issues and involving groups of scientists must be established first.

Finances of the magnitude necessary to sustain collaborative research on such a large scale do, however, demand more meticulous surveillance and management than do lesser individual endowments. Who accounts for expenditures? Who is held responsible in the event that affairs run amuck? In short, who is the watchdog?

The onus of administrative accountability falls upon the NIGMS-designated "principal investigator" of each Center. Correspondingly, each "principal" is free to delegate whatever administrative responsibility he or she feels is most advantageous to his or her specific program. The mood and tone of each Center is set by the "principal;" from that point, it evolves into a cooperative venture.

SCIENTIFIC RATIONALE

In the last several years, genetics research has stood at the brink of change. The magnificent portfolio of genetic information gleaned from lower organisms has set the stage for manipulative genetics in humans outside the scope of interdisciplinary research. What is required, what these Centers provide, are cores of geneticists surrounded by biochemists, molecular biologists, physicists and a myriad of other specialists.

The scientific rationale behind these Centers follows a logical progression from the simple to the complex, from what is already known to precisely what must be found out. Just as information derived from the scrutiny of lower organisms affords points of departure for research into higher organisms, these Centers propose to develop the most rudimentary genetic disorders as prototypes for the understanding and treatment of more complex genetic defects. This philosophy makes the ultimate goal of each Center a comprehension of polygenic defects, while the focus, at least for the moment, is on understanding the apparatus of Mendelian diseases.

Genetically related diseases may be categorized in order of ascending complexity according to the following scheme. The first and most elementary type of defect occurs when a single Mendelian gene is mutant and produces a disease such as hemophilia. The machinery responsible for the majority of Mendelian defects is a chemical alteration in the DNA causing an amino acid substitution in an end-product protein.

Chromosomal aberrations or abnormalities -- Turner's syndrome, for example -- represent the second category of disorders. "Here, generally," explains Dr. Arno Motulsky, principal investigator at the University of Washington Center, "the abnormality is not transmitted, but occurs as a *de nova* event in the sperm cells of the parents or in the first few divisions of the fertilized egg." As in the case of mongolism and other types of mental retardation, deleterious genetic material latent in the parents surfaces with profound consequences in the offspring.

The mechanisms of Mendelian defects and the pathways
through which chromosomal abnormalities produce multiple
clinical aberrations spearhead research at these seven
Centers. Much is presently known about the machinery of
single-gene diseases. The study of chromosomal pathways
is, in contrast, still in germinal stages. And though the
incidence and social consequences of these so-called Mendelian diseases seem monumental, they appear to shrink
against the backdrop of polygenic disorders.

The third or polygenic category of genetic defects is
that in which multiple genes interact to somehow cause an
individual to be more susceptible to a given disease.
Certain genes have, in fact, been implicated in a susceptibility to certain diseases. An entire spectrum of disorders ranging from atherosclerosis to heart disease to
diabetes and schizophrenia probably have such a mechanism
operative, according to Dr. Motulsky. "Case findings in
patients and particularly in their families, therefore,
have great public health significance."

Where funds are limited and resources finite, a multi-
million dollar research network, according to Dr. Bergmann,
cannot be justified on the basis of diseases that are comparatively "minor" in terms of their social impact. What
NIGMS appears to be banking on is that an understanding of
the lower members of the genetic disease hierarchy will
yield models for the treatment and cure of polygenic disorders.

THERAPEUTIC HIERARCHY

The shared and ultimate goal of each of these Centers
is the development of technologies by which the basic hereditary material DNA may be harnessed as therapeutic
agent for alleviating genetic diseases.

"If we can deliver the DNA that could cure or treat
some of these disorders," explains Dr. Vasken Aposhian of
the University of Maryland School of Medicine, "then the
mechanism of delivering that DNA could be useful in a
large number of diseases over the entire span of life-time
of various individuals." That same mechanism of delivery
-- "not what is inside, but the mechanism of delivery,"
stresses Dr. Aposhian -- may well be applicable to the
entire gamut of genetic disorders.

A broad spectrum of inborn metabolic errors eventuate
when faulty information in DNA orders up inadequate amounts
of a particular enzyme. Corrective, or at least therapeutic, strategy for these anomalies proceeds from the enzymatic reaction: substrate is converted to a product and
is catalyzed by an enzyme.

According to this reaction, treatment can be one or
more of the following procedures: (1) remove the accumulated substrate; (2) add product of the blocked reaction;
(3) add enzyme; (4) add the missing gene(s) which direct
synthesis of enzyme.

Therapies derived from this equation are in some

cases now available--- specifically in adding products or removing substrates. In other cases they remain largely a matter of optimistic speculation.

Measures one and two are, by and large, according to Dr. Bergmann, well documented. The treatment for hypopituitary dwarfism with growth hormone, for instance, is an example of the addition of a product in a blocked reaction. Another example is the use of insulin for the treatment of juvenile diabetes. In patients suffering from phenylketonuria, a prodigous volume of phenylalanine and phenylpyruvic acid accumulates. In such amounts, both of these compounds are, for reasons unknown, toxic. The condition can be alleviated through control of dietary intake thereby eliminating the accumulated substrate or cofactors.

"In a few instances," points out Dr. DeWitt Stetten, Jr., Director of NIGMS, "some of the symptoms of the diseases can be deferred or prevented by drastic dietary measures or by replacement of the end-product protein. But, none of these diseases can yet be cured in the sense that a victim no longer needs continuing treatment."

What is needed, what these Centers hope to construct, is a bridge between molecular biology and human biology. The addition of an enzyme to a system, or in other cases, of the missing gene or genes which direct synthesis of the enzyme that is deficient are the kingpins of such a bridge.

The use of enzymes in therapeutics has been comparatively unsuccessful in mammalian systems over the years, according to Dr. Aposhian. Several experiments with laboratory animals do, however, provide cases where enzyme catalase encased in plastic microcapsules permitted enzymes to diffuse gradually into cellular mediums with encouraging results. "Whether this will become a uniform method of treatment," explains Dr. Aposhian, "is of question because of the kinds of enzymes involved and the size of enzymes. It certainly is worthy of investigation." Researchers in at least two of the Genetics Centers -- the University of California Medical Center at San Francisco and the University of Texas Graduate School of Biomedical Sciences -- plan to do just that.

TRANSDUCTION AND TRANSFORMATION

The fourth, and most speculative and controversial method of treating inborn errors of metabolism -- addition of the missing gene(s) or "genetic engineering" -- may also prove itself to be the most promising. There is a need for DNA as a therapeutic agent because, says Dr. Aposhian, some of these diseases cannot be treated in any other way successfully.

How can a gene be introduced into a mammalian cell that is deficient in that gene? And assuming that the gene can be implanted, and that its genetic code corresponds to that which is deficient in the host, how can one guarantee that it will arrive at its intended destination, the liver for example, intact?

Explains Dr. Aposhian, "One thing we know from molecular biology is that we can give mammalian cell a piece of
DNA and have that DNA expressed genetically in its own
fashion if we protect that DNA by means of a protein coat.
We do have model systems of transformation of bacterial
systems, of infecting the bacterial cells with naked DNA
or DNA surrounded by something."

Prospects for a potential "mechanism of delivery"
stem from a phenomenon in molecular biology called transduction -- the delivery of a piece of host DNA inside a
virus to a cellular recipient. Model systems available
from successful implants in microbial genetics suggest
lanes of approach which may be applicable to mammalian
virology. Cause for optimism originates from experiments
which demonstrate that when mouse kidney cells are infected with polyoma virus, the resultant progeny display
evidence of total host DNA genome, as well as polyoma DNA.
Says Dr. Aposhian, "This was the first time that this type
of phenomena has been seen in animal virology. We think
we now may have something that will help us here."

The log of questions yet to be answered looms before
these Centers in well-defined but Herculean perspective.
Easy and reproducible methods for gene isolation, synthetic techniques for DNA manufacture and production, and
procedures to segregate and identify mammalian genes are
among the most pressing issues. Until further studies
substantiate that pseudovirus DNA is actually being physically integrated into host chromosomes (and not being
"chopped up" by enzymes in the cell), pseudovirus must be
administered at regular intervals, about every two to four
weeks. But what of antibody synthesis, the antigenicity
of the viral particle?

"What you eventually need for human gene therapy,"
speculated Dr. Aposhian in 1970, "if you jump now ten or
fifteen years or whatever it is, is human viruses that
have low antigenicity and are nonpathogenic."

PATIENTS AND PHYSICIANS

To what degree these seven Centers will entertain
diagnostic and consultative dialogue with patients and
physicians is only tacitly implied in the arithmetic of
these grants. Dr. Bergmann, when questioned regarding
guidlines for inpatient selection and treatment, explained
it thus: The mandate of these Centers is not to provide
health services, but to stimulate an environment for collaborative research. It all boils down to a question of
emphasis. Scientists working at these Centers are paid to
do research. There will, of course, be a coincident interaction between scientists and physicians. Diagnostic
consultation and patient counseling and treatment will
result, and that is all well and good. The point remains,
however, that these Centers were founded and exist as research facilities. Counseling and education are spin-off
products and are not necessarily basic to the design of

each Center. Specific Centers may elect to include these
as components of their programs. How integral a role counseling and education will play is up to these institutions.

There are instances where genetics research cannot
proceed exclusive of a patient population. Recognition
and diagnosis of carrier states, for example, involve both
patients and physicians through exhaustively detailed case
histories and on-site examinations.

"We are very interested," explains Dr. Margery Shaw,
principal investigator of the Medical Genetics Center
(MGC) at the University of Texas, "in studying individuals
who carry the gene but who don't manifest the disease."
To this end, "any" patient with genetic disease, and relatives, will be of paramount concern to the MGC clinical
unit.

Because NIGMS intends its allocations to fund research, only small monetary provisions have been earmarked
to defray hospitalization expenses incurred by patients.
Dr. Bergmann cautioned that such a measure is not meant
to discourage health services, but to underscore a research priority. NIGMS foresees most expenses being sub-
sidized through third party funds, such as Blue Cross,
Blue Shield, and other hospitalization insurance.

GENETIC CARTOGRAPHERS

The micro-worlds and mini-frontiers that genetic scientists will pioneer at these seven Centers pose vast un-
charted terrains. To discuss the scope and magnitude of
the genetic enigmas that each will explore is, according
to Dr. Bergmann, almost inadequate outside the context
of a 300-page thesis. What follows are profiles of experimental activity at these Centers. These sketches are presented as being representative of the type and nature of
research which will take place. They are not, in any case,
meant to cover the entire range of projects.

Chromosomal pathways remain a mystery that genetic
science has only recently begun to solve. At Johns Hopkins University, scientists will "map" the expression of
mammalian genes. Investigators hope to identify and align
genes on a linear map according to their individual characteristics and the frequency with which these characteristics are combined and expressed in offspring. As an adjunct to this study, researchers plan a study in population dynamics. Because genetic disorders manifest themselves more readily in inbred groups, genetic statisti-
cians will gather data from sects of Amish people, who
according to custom and religious conviction rarely wed
outside the boundaries of their immediate communities.

"Up until about one year ago," Dr. Bergmann explains,
"chromosomes could be distinguished only by shape. A
chromosome was fat, thin, in the shape of a cross, whatever." Today, with the advent of more sophisticated
staining and fluorescent procedures, scientists at the
University of Texas Center plan to specify chromosomes

according to banding techniques -- banding patterns are unique for each particular human chromosome. "I am very interested in the possibility that chromosomes have inherent banding patterns in different individuals," says Dr. Shaw.

Dr. William J. Schull, an associate of Dr. Shaw, will pursue demographic studies to ascertain the prevalence of genetic disease among Negroes, Whites and Chicanos in the Houston area. A corollary project proposes to determine any correlation which may exist between a propensity towards genetic disease and certain situations and environments. Through computer simulation, Dr. Schull will trace through several generations precisely what effects might result from various modes of family planning.

Three departments -- Pediatrics, Medicine and Bio-Chemistry -- at the University of California Medical Center at San Francisco will be directly involved in the Genetics Center there. Investigations will deal primarily with general and clinical problems related to chromosomal abnormalities and metabolic defects. Research will also be conducted on such basic problems as the mechanics of gene action, genetic control of metabolism and development, and the effects of genetic abnormalities on these functions. An important component of research there, according to Dr. Charles J. Epstein, principal investigator, will be the establishment of a centralized laboratory for growing cells in tissue culture, an asset important to both basic and clinical studies to be undertaken.

At the University of Washington School of Medicine in Seattle, scientists will investigate such diverse problems as the possible use of cyanate therapy for sickle cell anemia and the genetics of hyperkinetic children. Researchers also hope to develop more efficient technologies for pre-natal diagnosis and mass screening for inborn errors of metabolism.

Treatment of storage diseases through plasma transfusion, metabolic defects in red blood cells and hemoglobin variations, and hereditary factors in diabetes comprise the core of research problems to be studied at the Albert Einstein Genetics Center in New York City. Scientists there are also interested in mapping genes and chromosomes in an effort to determine data on translocation -- what enzymes are involved with what chromosomes. Dr. Bergmann terms this latter project "very basic to an understanding of genetics."

Myopia or near-sightedness intrigues researchers at the Mount Sinai School of Medicine in New York because of its primary incidence among children of Puerto Rican descent. "Studies of this sort," explains Dr. Bergmann, "point up the interdisciplinary nature of these Centers very well. To research myopia, for example, will require close collaboration with an optometrist."

The Procedures: Techniques of Genetic Engineering

Let us now turn our attention to some of the procedures that might be employed to achieve genetic manipulation of the human species. What procedures have been proposed to accomplish this end? Are any of these procedures currently available to us? How soon will other procedures be developed? How effective will these procedures be? What kinds of traits will science enable us to manipulate? What dangers are inherent in the use of genetic manipulation techniques? These and many related questions are of concern not only to scientific and medical professionals, but also to the general public. The readings included in this section of the book were selected because they provide at least tentative answers to some of these questions.

Bernard D. Davis' article, "Prospects for Genetic Intervention in Man," is a concise introduction to these questions and their possible answers. Davis suggests that curing single gene diseases (e.g., PKU) by replacement of defective genetic material with "correct" DNA à la transformation or transduction is a much more likely prospect than the genetic manipulation of behavioral traits which are controlled by many genes. Davis also briefly discusses cloning, predetermination of sex, and selective reproduction (breeding) as means of genetic engineering. The dangers of the various genetic manipulation procedures and their possible misuse for political and military purposes are briefly discussed. The article concludes with discussion of the damage that would be done to society if basic research were curtailed as a consequence of public apprehension over the possible misuse of genetic discoveries.

Nobel laureate, Joshua Lederberg, contrasts the high expectations held for genetic engineering with the reality of our present lack of ability to prevent and treat diseases having a genetic basis. We do not know how to prevent new mutations from occurring--indeed we usually do not even know that a recessive mutation has arisen until it is segregated in the homozygous condition, perhaps many generations later. In a few cases we can detect the presence of a deleterious recessive gene in heterozygous carriers and thus have a basis for counseling the carriers of the probability of their having a homozygous recessive child. We have amniocentesis which permits prenatal detection of a number of chromosome aberrations and a few enzyme deficiencies having a genetic basis. We can treat certain genetic diseases by replacing the missing gene product (e.g., insulin) or by removing from the diet substances (e.g., phenylalanine) which cannot be metabolized. On the whole, however, our arsenal of weapons against genetic disease is quite meager.

All of these facts, however, constitute a motivation for attempting to devise techniques of genetic manipulation--gene therapy via transformation or transduction and cloning being two examples. These facts also increase the risk that some of these genetic manipulation techniques will be used prematurely--before we are fully aware of the technical and ethical problems associated with them. In the case of cloning Lederberg specifically notes that "... we simply do not know enough about the question at either a technical or an ethical level... to dogmatize about whether or not it [cloning] should ever be done."

Lest the reader come to the conclusion that the development of all genetic engineering techniques lies in the future, he should note Davis' comment that "... any society wishing to direct the evolution of its gene pool already has available an alternative approach: selective breeding." The late Nobel laureate, H. J. Muller, in the third article in this section of the book, describes in some detail how such selective breeding could be accomplished through "voluntary choice of germ plasm." Essentially what Muller is advocating is AID--artificial insemination with the sperm of a donor. Moreover, Muller advocates that the donor be a person of established worth whose contributions to society and genetic superiority will have withstood the test of time. What is suggested, then, is that the semen of outstanding individuals be kept in the deep freeze of a sperm bank until such time that the individual's superiority is clearly recognized. The semen of such individuals could then be used to fertilize the eggs of numerous prospective mothers and thereby increase the frequency of the donor's superior genes in the human gene pool. Although written over a decade ago, Muller's article continues to have a futuristic ring to it suggesting that perhaps society in general is still not ready for even this currently feasible method of genetic manipulation.

In an article by Paul T. Libassi another tool currently available to medical science--prenatal diagnosis via amniocentesis is discussed. Some of the problems and limitations of amniocentesis and some refinements in chromosomal analysis are briefly described. Generally when amniocentesis is done, it is assumed that the prospective parents will agree to an abortion if the fetus is found to be genetically or cytologically defective. Thus, at the present time amniocentesis can be viewed as a technique useful in preventing the birth of a child with a chromosome aberration or one of a few biochemical defects. The use of amniocentesis has been quite limited (Libassi claims that some 1500 amniocenteses have been performed in the U.S. to date). How important amniocentesis might prove to be as a tool for changing gene and chromosome frequencies in the human gene pool remains to be seen.

Turning to techniques for genetic manipulation that might be widely used in the future, Friedmann and Roblin discuss the use of "...exogenous 'good' DNA... to replace the defective DNA in those [individuals] who suffer from genetic defects." Such gene therapy or gene surgery might be accomplished by means of supplying the defective human cells directly with the desired "good" DNA (transformation) or by having viruses serve as the vector for the introduction of the desired genetic material into the human genome (transduction). The authors propose ethico-scientific criteria which any prospective techniques for gene therapy should satisfy. The reader should note that where Muller was primarily concerned about modifying the human gene pool and thus the evolutionary future of man, Friedmann and Roblin direct their concern to alleviating genetic disease in individual patients, although they note that wide scale use of gene therapy would affect the human gene pool. Perhaps this difference in emphasis is fundamental as a basis for a decision as to whether or not genetic intervention is ethical: Is the intervention designed to correct a genetic defect of an individual, or is it designed to modify the evolutionary future of all mankind?

In a short article in Science Rabovsky reports on several recent attempts at transporting (via viruses) genes from cells of one species to cells of another. The experiments of Merril et al., for example, show that a bacterial gene can be incorporated into a human cell by a phage, and that the bacterial gene can supply an enzyme missing in the human cell. Although done in vitro, such experiments suggest that it may not be long before methods will be found for transducing genes in humans with single gene defects.

Another technique which could permit genetic engineering is clonal or asexual reproduction. Nobelist James D. Watson discusses the possible use of this technique in a stimulating article from Intellectual Digest. Although accomplished only with amphibians to date, Watson predicts the cloning of mice in the not-too-distant future. This will pave the way for cloning other mammals, including man. Watson cites good reasons why basic research in related areas of cell biology should not be curtailed and suggests the possibility of an international agreement specifically declaring the illegality of human cloning.

In "Genetic Manipulation and Man," Darrel English provides us with a good summary discussion of a number of techniques that might be used to achieve genetic engineering. He discusses these in the historical perspective of the eugenics movement of the late nineteenth and early twentieth centuries. English concludes his article with brief comments on the ethical implications of genetic engineering.

Kurt Hirschhorn's "On Re-Doing Man" provides the reader with historical insight into the eugenics movement, contrasting positive eugenics with negative eugenics and discussing the role of euthenics in allowing the genetically defective individual to develop and live normally. Methods of genetic manipulation are briefly mentioned and the article concludes with a discussion of ethical considerations relative to genetic engineering. Dr. Hirschhorn's article was included as the final selection in this part of the book because its content serves as a "bridge" connecting the discussion of techniques of genetic manipulation with the papers in the next section of the book which deal primarily with moral, ethical and legal implications of genetic engineering.

Prospects for Genetic
Intervention in Man

Bernard D. Davis

Extrapolating from the spectacular successes of molecular genetics, a number of essays and symposia (*1*) have considered the feasibility of various forms of genetic intervention (*2*) in man. Some of these statements, and many articles in the popular press, have tended toward exuberant, Promethean predictions of unlimited control and have led the public to expect the blueprinting of human personalities. Most geneticists, however, have had more restrained second thoughts.

Nevertheless, recent alarms about this problem have caused wide public concern, and understandably so. With nuclear energy threatening global catastrophe and with so many other technological advances visibly damaging the quality of life, who would wish to have scientists tampering with man's inner nature? Indeed, fear of such manipulation may arouse even more anxiety than fear of death. The mass media have accordingly welcomed sensational pronouncements about the dangers.

While such dangers clearly exist, it also seems clear that some scientists have dramatized them (*3*) in order to help persuade the public of the need for radical changes in our form of government (*4*). But however laudable the desire to improve our social structure, and however urgent the need to improve our protection against harmful uses of science and technology, exaggeration of the dangers from genetics will inevitably contribute to an already

SCIENCE, 1970, Vol. 170, pp. 1279-1283. Copyright 1970 by the American Association for the Advancement of Science.

distorted public view, which increasingly blames science for our problems and ignores its contributions to our welfare. Indeed, irresponsible hyperbole on the genetic issue has already influenced the funding of research (5). It therefore seems important to try to assess objectively the prospects for modifying the pattern of genes of a human being by various means. But let us first note two genetic principles that must be taken into account.

Relevant Genetic Principles

Polygenic traits and behavioral genetics. The recognition of a gene, in classical genetics, depends on following the distribution of two alternative forms (alleles) from parents to progeny. In the early years of genetics, after the rediscovery of Mendel's laws in 1900, this analysis was possible only for those genes that exerted an all-or-none control over a corresponding monogenic trait—for example, flower color, eye color, or a hereditary disease such as hemophilia. The study of such genes has continued to dominate genetics. However, monogenic traits constitute a small, special class. Most traits are polygenic: that is, they depend on multiple genes, and so they vary continuously rather than in an all-or-none manner. Moreover, each gene itself is polymorphic—that is, it is capable of existing, as a result of mutation, in a variety of different forms (alleles); and though the protein products of these alleles differ only slightly in structure, they often differ markedly in activity.

For our purpose it is especially pertinent that the most interesting human traits—relating to intelligence, temperament, and physical structure—are highly polygenic. Indeed, man un-

doubtedly has hundreds of thousands of genes for polygenic traits, compared with a few hundred recognizable through their control over monogenic traits. However, the study of polygenic inheritance is still primitive; and the difference from monogenic inheritance has received little public attention. Education on the distinction between monogenic and polygenic inheritance is clearly important if the public is to distinguish between realistic and wild projections for future developments in genetic intervention in man.

Interaction of heredity and environment. The study of polygenic inheritance is difficult in part because it requires statistical analysis of the consequences of reassortment, among the progeny, of many interacting genes. In addition, even a full set of relevant genes does not fixedly determine the corresponding trait. Rather, most genes contribute to determining a *range of potential* for a given trait in an individual, while his past and present environments determine his phenotype (that is, his actual state) within that range. At a molecular level the explanation is now clear: the structure of a gene determines the structure of a corresponding protein, while the interaction of the gene with subtle regulatory mechanisms, which respond to stimuli from the environment, determines the amount of the protein made. Hence, the ancient formulation of the question of heredity versus environment (nature versus nurture) in qualitative terms has presented a false dichotomy, which has led only to sterile arguments.

Possibilities in Genetic Manipulation

Somatic cell alteration. Bacterial genes can already be isolated (6) and synthesized (7); and while the isolation of human genes still appears to be a formidable task, it may also be accomplished quite soon. We would then be able to synthesize and to modify human genes in the test tube. However, the incorporation of externally supplied genes into human cells is another matter. For while small blocs of genes can be introduced in bacteria, either as naked DNA (transformation) or as part of a nonlethal virus (transduction), we have no basis for estimating how hard it will be to overcome the obstacles to applying these methods to human cells. And if it does become possible to incorporate a desired gene into some cells, in the intact body, incorporation into all the cells that could profit thereby may well remain difficult. It thus seems possible that diseases depending on deficiency of an extracellular product, such as insulin, may be curable long before the bulk of hereditary diseases, where an externally supplied gene can benefit only those defective cells that have incorporated it and can then make the missing cell component.

Such a one-shot cure of a hereditary disease, if possible, would clearly be a major improvement over the current practice of continually supplying a missing gene product, such as insulin. (It could be argued that improving the soma in this way, without altering the germ cells, would help perpetuate hereditary defectives; but so does conventional medical therapy.) The danger of undesired side effects, of course, would have to be evaluated, and the day-to-day medical use of such material would have to be regulated: but these problems do not seem to differ significantly from those encountered with any novel therapeutic agent.

Germ cell alteration. Germ cells may prove more amenable than somatic cells to the introduction of DNA, since they could be exposed in the test tube and therefore in a more uniform and controllable manner. Another conceivable approach might be that of *directed mutagenesis*: the use of agents that would bring about a specific desired alteration in the DNA, such as reversal of a mutation that had made a gene defective. So far, however, efforts to find such directive agents have not been successful: all known mutagenic agents cause virtually random mutations, of which the vast majority are harmful rather than helpful. Indeed, before a mutagen could be directed to a particular site it would probably have to be attached first to a molecule that could selectively recognize a particular stretch of DNA (8); hence a highly selective mutagen would have to be at least as complex as the material required for selective genetic recombination.

If predictable genetic alteration of germ cells should become possible it would be even more useful than somatic cure of monogenic diseases, for it could allow an individual with a defective gene to generate his own progeny without condemning them to inherit that gene. Moreover, there would be a long-term evolutionary advantage, since not only the immediate product of the correction but also subsequent generations would be free of the disease.

Genetic modification of behavior. In contrast to the cure of specific monogenic diseases, improvement of the highly polygenic behavioral traits would almost certainly require the replacement, in germ cells, of a large but specific complement of DNA. Since I find such replacement, in a controlled manner, very hard to imagine, I suspect that such modifications will remain indefinitely in the realm of science fiction, like the currently popular extrapolation from the transplantation of a kidney or a heart, with a few tubular connections, to that of a brain, with hundreds of thousands of specific neural connections. However, this consideration would not apply to the possibility of impairing cerebral function by genetic transfer, since certain monogenic diseases are known to cause such impairment.

Copying by asexual reproduction (cloning). We now know that all the differentiated somatic cells of an animal (those from muscle, skin, and the like) contain, in their nuclei, the same complete set of genes. Every somatic cell thus contains all the genetic information required for copying the whole organism. In different cells different subsets of genes are active, while the remainder are inactive. Accordingly, if it should become possible to reverse the regulatory mechanism responsible for this differentiation any cell could be used to start an embryo. The individual could then be developed in the uterus of a foster mother, or eventually in a glorified test tube, and would be an exact genetic copy of its single parent. Such asexual reproduction could thus be used to produce individuals of strictly predictable genetic endowment; and there would be no theoretical limit to the size of the resulting clone (that is, the set of identical individuals derivable from a single parent and from successive generations of copies).

Though differentiation is completely reversible in the cells of plants (as in the transfer of cuttings), it is ordinarily quite irreversible in the cells of higher animals. This stability, however, depends on the interaction of the nucleus with the surrounding cytoplasm; and it is now possible to transfer a nucleus, by microsurgery or cell fusion, into the cytoplasm of a different kind of cell. Indeed, in frogs differentiation has been completely reversed in this way: when the nucleus of an egg cell is replaced by a nucleus from an intestinal cell embryonic development of the hybrid cell can produce a genetic replica of the donor of the nucleus (9). This result will probably also be accomplished, and perhaps quite soon, with cells from mammals. Indeed, there is considerable economic incentive to achieve this goal, since the copying of champion livestock could substantially increase food production.

Another type of cloning can already be accomplished in mammals: when the relatively undifferentiated cells of an early mouse embryo are gently separated each can be used to start a new embryo (10). A large set of identical twins can thus be produced. However, they would be copies of an embryo of undetermined genetic structure, rather than of an already known adult. This procedure therefore does not seem tempting in man, unless the production of identical twins (or of greater multiplets) should develop special social values, such as those suggested by Aldous Huxley in *Brave New World*.

Predetermination of sex. Though no one has yet succeeded in directly controlling sex by separating XX and XY sperm cells, this technical problem should be soluble. Moreover, in principle it is already possible to achieve the same objective indirectly by aborting embryos of the undesired sex: for the sex of the embryo can be diagnosed by tapping the amniotic fluid (amniocentesis) and examining the cells released into that fluid by the embryo.

Wide use of either method might cause a marked imbalance in the sex ratio in the population, which could lead to changes in our present family structure (and might even be welcomed in a world suffering from overpopulation). Alternatively, new social or legal pressures might be developed to avert a threatened imbalance *(11)*. But though there would obviously be novel social problems, I do not think they would strain our powers of social adaptation nearly as much as some urgent present problems.

Selective reproduction. A discussion of the prospects for molecular and cellular intervention in human heredity would be incomplete without noting that any society wishing to direct the evolution of its gene pool already has available an alternative approach: selective breeding. This application of classical, transmission genetics has been used empirically since Neolithic times, not only in animal husbandry, but also, in various ways (for example, polygamy, *droit de seigneur*, caste system), in certain human cultures. Declaring a moratorium on genetic research, in order to forestall possible future control of our gene pool, would therefore be locking the barn after the horse was stolen.

Having reviewed various technical possibilities, I would now like to comment on the dangers that might be presented by their fulfillment and to compare these with the consequences of efforts to prevent this development.

Evaluation of the Dangers

Gene transfer. I have presented the view that if we eventually develop the ability to incorporate genes into human germ cells, and thus to repair monogenic defects, we would still be far from specifying highly polygenic behavioral traits. And with somatic cells such an influence seems altogether excluded. For though genes undoubtedly direct in considerable detail the pattern of development of the brain, with its network of connections of 10 billion or more nerve cells, the introduction of new DNA following this development clearly could not redirect the already formed network; neither could we expect it to modify the effect of learning on brain function.

To be sure, since we as yet have little firm knowledge of behavioral genetics we cannot exclude the possibility that a few key genes might play an especially large role in determining various intellectual or artistic potentials or emotional patterns. But even if it should turn out to be technically possible to tailor the psyche significantly by the exchange of a small number of genes in germ cells, it seems extremely improbable that this procedure would be put to practical use. For it will always be much easier, as Lederberg *(12)* has emphasized, to obtain almost any desired genetic pattern by copying from the enormous store already displayed in nature's catalog.

While the improvement of cerebral function by polygenic transfer thus seems extremely unlikely, one cannot so readily exclude the technical possibility of impairing this function by transfer of a monogenic defect. And having seen genocide in Germany and massive defoliation in Vietnam, we can hardly assume that a high level of civilization provides a guarantee against such an evil use of science. However, several considerations argue against the likelihood that such a future technical possibility would be converted into reality. The most important is that monogenic diseases, involving hormonal imbalance or enzymatic deficiencies, produce gross behavioral defects, whose usefulness to a tyrant is hard to imagine. Moreover, even if gene transfer is achieved in cooperating individuals, an enormous social effort would still be required to extend it, for political or military purposes, to mass populations. Finally, in contrast to the development of nuclear energy, which arose as an extension of already accepted military practices, the potential medical value of gene transfer is much more evident than its military value; hence a "genetic bomb" could hardly be sprung on the public as a secret weapon. Accordingly, we are under no moral obligation to sacrifice genetic advances now in order to forestall such remote dangers: if and when gene transfer in man becomes a reality there would still be time to assert the cultural and medical traditions that would promote its beneficial use and oppose its abuse.

This last obstacle would be eliminated if it should prove possible to develop a virus that could be used to infect a population secretly with specific genes, and it is the prospect of this ultimate horror that seems to cause most concern. However, for reasons that I have presented above the technical possibility of producing useful modifications of personality by infections of germ cells seems extremely remote, and the possibility of doing so by infecting somatic cells in an already developed individual seems altogether excluded. These fears thus do not seem realistic enough to help guide present policy. Nevertheless, the problem cannot be entirely ignored: in a country that has recently been embarrassed by its accumulation of rockets containing nerve gas even the remote possibility of handing viral toys to Dr. Strangelove will require vigilance.

Genetic copies. If the cloning of mammals becomes technically feasible its extension to man will undoubtedly be very tempting, on the grounds that enrichment for proved talent by this means might enormously enhance our culture, while the risk of harm seemed small. Since society may be faced with the need to make decisions in this area quite soon, I would like to offer a few comments in the hope of encouraging public discussion.

On the one hand, in fields such as mathematics or music, where major achievements are restricted to a few especially gifted people, an increase in their number might be enormously beneficial—either as a continuous supply from one generation to another or as an expanded supply within a generation. On the other hand, a succession of identical geniuses might exert an excessively conservative influence, depriving society of the richness that comes from our inexhaustible supply of new combinations of genes. Or genius might fail to flower, if its drive depended heavily on parental influence or on cultural climate. And in the literary, social, and political areas the cultural climate surely plays so large a role that there may be little basis for expecting outstanding achievement to be continued by a scion. The world might thus be quite disappointed by the contributions of another Tolstoy, Churchill, or Martin Luther King, or even another Newton or Mozart. Moreover, though experience with monozygotic twins is somewhat reassuring, persons produced by copying might suffer from a novel kind of "identity crisis."

Though our system of values clearly places us under moral obligation to do everything possible to cure disease, there is no comparable basis for using cloning to advance culture. The responsibility for initiating such a radical departure in human reproduction would be grave, and surely many will feel that we should not do so. But I suspect that it would be impossible to enforce any such prohibition completely: the potential gain seems too large, and the procedure would require the cooperation of only a very small group of people. Hence whatever the initial social concensus, I suspect that a stable attitude would not emerge until after some early tests, whether legal or illegal, had demonstrated the magnitude of the problems and of the gains.

A much greater threat, I believe, would be the use of cloning for the large-scale amplification of a few selected individuals. Who would wish to send a child to a school with a large set of identical twins as his classmates? Moreover, the success of a species depends not only on its adaptation to its present environment but also on its possession of sufficient genetic variety to include some individuals who could survive in any future environment. Hence if cloning were extended to the point of markedly homogenizing the population, it could create an evolutionary danger. However, we have already lived for a long time with a similar possibility: any male can provide a virtually limitless supply of germ cells, which can be used in artificial insemination; yet genetic homogenization by this means has not become the slightest threat. Since cloning is unlikely to become nearly so easy it is difficult to see a rational basis for the fear that its technical possibility would increase the threat.

Implications for genetic research. Though the dangers from genetics seem to me very small compared with the immense potential benefits, they do exist: its applications could conceivably be used unwisely and even malevolently. But such potential abuses cannot be prevented by curtailing genetic research. For one thing, we already have on hand a powerful tool

(selective breeding) that could be used to influence the human gene pool, and this technique could be used as wisely or unwisely as any future additional techniques. Moreover, since the greatest fear is that some tyrant might use genetic tools to regulate behavior, and especially to depress human potential, it is important to note that we already have on hand pharmacological, surgical, nutritional, and psychological methods that could generate parallel problems much sooner. Clearly, we shall have to struggle, in a crowded and unsettled world, to prevent such a horrifying misuse of science and to preserve and promote the ideal of universal human dignity. If we succeed in developing suitable controls we can expect to apply them to any later developments in genetics. If we fail—as we may—limitations on the progress of genetics will not help.

If, in panic, our society should curtail fundamental genetic research, we would pay a huge price. We would slow our current progress in recognizing defective genes and preventing their spread; and we would block the possibility of learning to repair genetic defects. The sacrifice would be even greater in the field of cancer: for we are on the threshold of a revolutionary improvement in the control of these malignant hereditary changes in somatic cells, and this achievement will depend on the same fundamental research that also contributes toward the possibilities of cloning and of gene transfer in man. Finally, it is hardly necessary to note the long and continuing record of nonmedical benefits from genetics, including increased production and improved quality of livestock and crops, steadier production based on resistance to infections, vastly increased yields in antibiotic and other industrial fermentations, and, far from least, the pride that mankind can feel in one of its most imaginative and creative cultural achievements: understanding of some of the most fundamental aspects of our own physical nature and that of the living world around us.

While specific curtailment of genetic research thus seems impossible to justify, we should also consider briefly the broader proposal (see, for example, 8) that we may have to limit the rate of progress of science in general, if we wish to prevent new powers from developing faster than an inadequate institutional framework can be adjusted to handle them. While one can hardly deny that this argument may be valid in the abstract, its application to our present situation seems to me dangerous. No basis is yet in sight for calculating an optimal rate of scientific advance. Moreover, only recently have we become generally aware of the need to assess and control the true social and environmental costs of various uses of technology. Recognition of a problem is the first step toward its solution, and now that we have taken this step it would seem reasonable to assume, until proved otherwise, that further scientific advance can contribute to the solutions faster than it will expand the problems.

Another consideration is that we cannot destroy the knowledge we already have, despite its potential for abuse. Nor can we unlearn the scientific method, which is available for all who wish to wrest secrets from nature. So if we should choose to curtail research in various fundamental areas, out of fear of possible long-range application, we must recognize that other societies may make a different choice. Knowledge is power, and power can be used for good or for evil; and, since the genie that brings new knowledge is already out of the bottle, we must learn to direct the use of the resulting power rather than curse the genie or try to confine him.

We cannot see how far the use of science as a scapegoat for many of our social problems will extend. But the gravity of the threat may be underscored by recalling that another politically based attack on science, Lysenkoism, utterly destroyed genetics in the Soviet Union and seriously crippled agriculture, from 1935 to 1965 (13). [This development illustrates ironically the unstable relation between political and scientific ideas: for Karl Marx had unsuccessfully requested permission to dedicate the second volume of *Das Kapital* to Charles Darwin (14)!] Moreover, the current attacks on genetics from the New Left can build on, and have no doubt contributed to, widespread public anxiety concerning gene technology. Thus while a recent report prepared for the American Friends Service Committee (15) presents an open and thoughtful view on such questions as contraception, abortion, and prolongation of the period of dying, it is altogether opposed to any attempted genetic intervention, including the cure of hereditary disease.

Genetics will surely survive the current attacks, just as it survived attacks from the Communist Party in Moscow and from fundamentalists in Tennessee. But meanwhile if we wish to avert the danger of some degree of Lysenkoism in our country we may have to defend vigorously the value of objective and verifiable knowledge, especially when it comes into conflict with political, theological, or sociological dogmas.

References and Notes

1. P. B. Medawar, *The Future of Man* (Basic Books, New York, 1960); Symposium on "Evolution and Man's Progress," *Daedalus*

(Summer, 1961); G. Wolstenholme, Ed., *Man and His Future* (Little, Brown, Boston, 1963); J. Lederberg, *Nature* 198, 428 (1963); J. S. Huxley, *Essays of a Humanist* (Harper and Row, New York, 1964); T. M. Sonneborn, Ed., *The Control of Human Heredity and Evolution* (Macmillan, New York, 1965); R. D. Hotchkiss, *J. Hered.* 56, 197 (1965); J. D. Roslansky, Ed., *Genetics and the Future of Man* (Appleton-Century-Crofts, New York, 1966); N. H. Horowitz, *Perspect. Biol. Med.* 9, 349 (1966).

2. The term "genetic engineering" seemed at first to be a convenient designation for applied molecular and cellular genetics. However, I agree with J. Lederberg [*The New York Times*, Letters to the editor, 26 September (1970)] that the overtones of this phrase are undesirable.

3. Editorials, *Nature* 224, 834, 1241 (1969); J. Shapiro, L. Eron, J. Beckwith, *ibid.*, p. 1337.

4. J. Beckwith, *Bacteriol. Rev.* 34, 222 (1970).

5. P. Handler, *Fed. Proc.* 29, 1089 (1970).

6. J. Shapiro, L. MacHattie, L. Eron, G. Ihler, K. Ippen, J. Beckwith, *Nature* 224, 768 (1969).

7. K. L. Agarwal, and 12 others, *ibid.* 227, 27 (1970).

8. S. E. Luria, in *The Control of Human Heredity and Evolution*, T. M. Sonneborn, Ed. (Macmillan, New York, 1965), p. 1.

9. R. Briggs and T. J. King, in *The Cell*, J. Brachet and A. E. Mirsky, Eds. (Academic Press, New York, 1959), vol. 1; J. B. Gurdon and H. R. Woodward, *Biol. Rev.* 43, 244 (1968).

10. B. Mintz, *J. Exp. Zool.* 157, 85, 273 (1964).

11. A. Etzioni, *Science* 161, 1107 (1968).

12. J. Lederberg, *Amer. Natur.* 100, 519 (1966).

13. Z. A. Medvedev, *The Rise and Fall of T. D. Lysenko* (Columbia Univ. Press, New York, 1969).

14. T. Dobzhansky, *Mankind Evolving* (Yale Univ. Press, 1962), p. 132.

15. *Who Shall Live?* Report prepared for the American Friends Service Committee (Hill and Wang, New York, 1970).

GENETIC ENGINEERING AND THE AMELIORATION OF GENETIC DEFECT

Joshua Lederberg

Few subjects pose as many difficulties for rational discussion as does the bearing of genetic research on human welfare. It is monotonously coupled with such inflammatory themes as racism, the decline of the species, overpopulation, hidden genocide, religious debates on abortion and contraception, the plight of the individual in mass society, and "how many generations of idiots is enough?" Apart from these cliches of social controversy, we also face serious problems in clarifying the technical context in a field where scientific innovations have far outpaced their technical application and applicability to man. Should we spend much time worrying about the ethical implications of the genetic findings of the next century, when we must do this on the basis of a set of assumptions about the human conditions that will surely change dramatically in every other way?

The gap between elementary principle and practical realization poses many dilemmas for a fair-minded evaluation, and a brief commentary cannot do justice to all of the relevant issues. The scientist who concentrates on exposing the technical possibilities will be castigated for ignoring the ethical implications; and if he goes further afield, he will be accused of sermonizing, and indeed may be overreaching his particular area of competence.

BIOSCIENCE, 1970, Vol. 20, pp. 1307-1310.

According to journalistic accounts, we will shortly be writing prescriptions for human quality to order. "Do you want your baby to be 8 ft tall, or have four hands—just tell the geneticist, and he will arrange it for you," goes this line of advertisement. But the most sophisticated geneticist today is baffled by challenges such as Huntingdon's disease. Will the son of an afflicted father be afflicted later in life? What can he do to assure that his own children will not have it?

Perhaps some year soon we will know enough at least to recognize the genotype before neuronal degeneration has been irreversibly set in motion. But our failure to be able to provide significant help today is a humbling reality next to the effusive but justifiable predictions about future accomplishments.

What then of the bold claims for a brave new world of genetic manipulation? Their substance is grounded on the recent solution of many fundamental mysteries of genetic biochemistry. Many of the obstacles to genetic engineering, apart from the moral and political questions that it may pose, are technological; which is to say that their solution is consistent with our basic scientific knowledge of the gene. But this is as if to say that "merely technical obstacles" prevent building a land bridge from San Francisco to Honolulu. It is safe to predict that this enterprise will never eventuate, not merely because it would be a million times more costly than previous bridges, but rather because other challenges will compete for the energies and resources. The presumed benefits will be achieved by other routes, the image of the trans-Pacific bridge will persist as a metaphor, reminding us of technical achievements in other fields of transportation and communication and of the political prodigy of the evolution of a specific island from dependency to statehood.

Construction works, like bridges, are open to evaluation and judgment obvious to all. Biotechnical projects are more likely to be cloaked in an esoteric jargon that depletes common sense. We may then hear the most absurd generalizations, such as "whatever is technically feasible tends to get done."

Anyone who has actually labored to do *anything* knows that the more appropriate slogans are "almost nothing ever gets done, especially if it costs money." Or "when a need is generally perceived, articulately formulated and wisely analyzed, the technical problems will be surmounted. But this will happen much sooner if a mass advertising campaign can be built around it." If transoceanic bridges become fashionable (as might happen incidentally during an arms race), they will be built, and probably the same can be said for genetic engineering. Our foresight about the future will prove to be right or wrong more on the basis of the predictability of fashion than of the scientific bases tò technical solutions. I do not venture to foresee the directions that such fashion may take. This is more the realm of the political theorists or of authors like John Brunner *(Stand on Zanzibar)*.

Where then does the scientist fit into such a discussion?

He can fairly justify his life and work in terms of fundamental knowledge about nature. Studies on the implantation of nuclei into eggs of different genotypes are a rewarding approach to learning how genes function and how this relates to the development of the egg. Were they done for the purported purpose of learning the technology of cloning in man, we would then be obliged to set a priority (positive or negative) on it from the standpoint of the human values that might justify or repudiate the investment. A small amount of "scientific" effort is, unhappily, biased by the expectation of the publicity that will attach to spectacular demonstrations of "behavior control" or "gene control" (shall we also say "moon-walking?") for its own sake; the scientific community can seek to impose criteria of *scientific* validity for the funding of such projects; or, failing this, to dissociate itself from the responsibility when (as is customary) it does not have the authority to make the critical decisions.

Alternatively, the scientist can function as the actual or effective member of a technological team that will address itself to the solution of grave problems that encumber human welfare. Then we must (and usually do) insist that the problems are real ones, and that technical solutions are credible. What is more often obscured is the need to examine all the side effects, to inhibit the premature exploitation of new cures that may be far worse than the disease, to assure that as much sophistication goes into looking for the side effects as was eagerly purchased for the primary solution. The hazards of suboptimal solutions are well appreciated for drugs, but we are just now feeling their full force in such disparate fields as pesticides and auto transport. Pesticide poisoning and air pollution have been figured as technological jinn. It would be more fair to lay the blame on technological idiocy—the refusal to make the economic investments needed to develop all the science required for the safe and healthful utilization of the new tricks.

What then are the *problems* to which genetic science can be applied? Some may think of rescuing man from the prospect of nuclear annihilation by recasting the genes for aggression, or acquiescence, that are supposed to predestine a future of territorial conflict. Even if we postulate for the sake of agreement that we knew the genetics of militarism, we have no way to apply it without solving the political problem that is the primary difficulty. If we could agree upon applying genetic (or any other effective) remedies to global problems in the first place, we probably would need no recourse to them in the actual event.

The converse argument applies to the gloomier predictions of totalitarian **abuse of a genetic technology. The scenario of Brave New World is well advertised by now, and no one doubts that a modern slave state would reinforce its class stratification by genetic controls. But it could not do so without having instituted slavery in the first place, for which the control of the mass media presents much more immediate dangers than knowledge of DNA. It is indeed true that I might fear the control of my behavior through electrical impulses directed into my brain, but (possibly excepting television) I do not accept the implantation of the electrodes except at the point of a gun, and this is the problem.**

So much for the grand designs of genetic engineering. There remain the very real tragedies of genetic disease. The societal interest in preventing or ameliorating mental retardation and other forms of congenital malformation is obvious. (The true cost of lifetime maintenance of a 21-trisomy approaches a megadollar.) It is also entirely congruent with the needs of the family, and, if we believe in the nobility of man and the worth of *human* life, also of the afflicted child as well.

The most effective avenues of preventing genetic disease include (1) the primary prevention of gene mutations, and (2) the detection and humane containment of the DNA lesions once introduced into the gene pool. The "natural" mutation process in man results in the introduction of a new bit of genetic misinformation once in every ten gametes. Most of the human cost of this "mutational load" is paid during early stages of fertilization and pregnancy, where it makes up a fair part of the total fetal wastage. But about 2% of newborns suffer from a recognizable discrete genetic defect. This is just the tip of the iceberg; the heritability of many common diseases suggests that from one-quarter to one-half of *all* disease is of genetic origin, for there are important variations in susceptibility to the frankest of environmental insults.

Not all of this health deficit can be attributed to recurrent mutations. An unknown proportion results from the selective advantage that is paradoxically associated with the heterozygous state of many genes, even some with lethal effect in the homozygote (like sickle-cell hemoglobin). Nevertheless, a significant part of medicine—much more than most practitioners overtly recognize—is in fact directed to lesions that **are inherently preventable if we could control the mutation process in the background.**

About a tenth of the "natural mutation rate" can be attributed to background radiation from cosmic rays and from radioactive potassium and other isotopes in our natural environment. Therefore, doubling the background, which would correspond to the maximum permissible standards now advocated by federal agenices, would add another 10% to the existing mutation rate: 1/9 rather than 1/10 of our gametes would carry deleterious mutations. This is an enormous impact in absolute terms, a modest increase in relative terms. We must nevertheless pay careful attention to the benefits that would be connected with this level of radiation exposure to be sure we are getting a fair bargain.

It must be pointed out that industrial nuclear energy activities today add less than 10% to the average radiation background (hence less than 1% to the mutation rate); medical X-rays add 50% and 5% respectively. The more prevalent standards for the judicious and cost-effective use of diagnostic X-rays do not necessarily or automatically excuse the dispensable residue.

A significant portion of "spontaneous" mutations must be attributed to environmental chemicals, many of which are clearly established as mutagens in laboratory experiments (for example, the peroxy-compounds that characterize smog). The extent to which such materials reach the germ cells is absolutely unknown at present. However, there are good reasons to believe that (1) the induction of mutations, in germ cells, and of cancer, in somatic cells, are fundamentally similar processes—most chemical carcinogens being also mutagenic when properly tested; and (2) a large part of the incidence of cancer is of chemical-environmental origin, cigarette-smoking being only the best-known and best-advertised example. It therefore follows that environmental mutagenesis is equally prevalent, if the relative effects of radiation in the two systems are any hint, the cryptic penalties of the mutations are likely, in the long run, to exact the larger price in human misery.

The direct observation of human populations for evidence of changes in mutation rates is an almost hopeless task. We have no way of managing the tangle of known and unknown environmental influences that bear on different individuals. Nor do we have tractable assays for the occurrence of new mutations, whose manifestation may be delayed (by transmission through heterozygotes) for many generations, or confused with malformations due to pre-existing mutant genes, or to non-genetic causes. If we had to rely upon epidemiological evidence, we would still lack persuasive evidence that radiation, even in barely sublethal doses, was mutagenic in man.

Our only recourse is the laboratory experiment with convenient mammals such as the mice, and sometimes even more efficiently with viruses and microorganisms. Even so, only the most potent mutagens can be identified with mice, and many uncertainties will remain that cannot be resolved given possible differences in metabolism and transport, cell selection, intrinsic sensitivity, and the duration and style of life of the human versus the experimental species. Very recently, we have been able to look deeper into the mechanism of chemical mutation, at the level of the structure and repair of the DNA molecule, and new procedures may be developed that can give us more reliable information on the susceptibility of the DNA of human cells to environmental insults. They may also give us clues to ways of neutralizing mutational lesions, either by blocking the primary effects of mutagens on DNA, or by bolstering the natural mechanisms of "editing" and repair of DNA information.

Once a mutation has been allowed to occur in a gamete, and this then participates in fertilization and the production of a new individual, we face a much more difficult problem in any effort at genetic hygiene. For now we must deal with the destinies of human

75

individuals, not merely the chemistry of an isolated segment of DNA. Our problem, seen in the whole, is compounded by every humanitarian effort to compensate for a genetic defect, insofar as this shelters the carrier from natural selection. So it must be accepted that medicine, even prenatal care (which may permit the fragile fetus to survive), already intrudes on the question "Who shall live?", the challenge so often thrust at rational discussions of policies that might influence the frequencies of deleterious genes. It is so difficult to do only good in such matters that we are better off putting our strongest efforts into the prevention of mutations, so as to minimize the heavy moral and other burdens of decision once the gene pool has been seeded with them.

We still cannot evade an evolutionary legacy of genetic damage that would remain with us for generations, even if all new mutation could be stopped by fiat. Our fundamental resources remain very feeble. In a few cases, we can diagnose the heterozygous carriers of recessive mutations, and the genetic counselor can then advise the prospective parents of the odds that they will have affected children. Where voluntary childlessness is unacceptable, it is also sometimes possible to monitor a pregnancy by sampling cells from the amniotic fluid. This can enable the mother to proceed with confidence or to request an elective abortion on the basis of firm knowledge of the genotype of the fetus. We can expect a rapid extension of technical facilities for such diagnoses. At present, they are limited to examination of the chromosomes (for gross chromosomal abnormalities, like Downs' syndrome), and to enzyme assays on cultured cells, which can diagnose a few dozen rare diseases with varying degrees of reliability. We will surely be learning during the next decade how to use much more sophisticated approaches to the structure of the DNA and RNA of such cells for more basic diagnostic methods.

In many cases, a deeper understanding of the casual chain by which a DNA alteration leads to pathology may help us devise new forms of therapy to compensate for the genetic defect. This may be as crude as the use of insulin in diabetes or as subtle as the use of controlled diets in phenylketonuria. (Both approaches are valuable; neither entirely satisfactory.)

Another approach to constructive therapy, which may mitigate a variety of diseases, is an extension of the existing uses of specific virus strains. At present, their role in medicine is confined to their use as vaccines, for the provocation of immunity against related, wild viruses. This is a specialized example of the modification of cell metabolism by inoculated DNA, discovered empirically by Jenner, and still quite imperfectly understood (our ignorance being concealed by the conceptualizations of clinical virology, which still fail to explain just how a vaccine works—e.g., to state just which cells of the vaccinated individual are carrying the viral genetic information, and in what form. We can visualize the engineering of other viruses so that they will introduce compensatory genetic information into the appropriate somatic cells to restore functions that are blanked out in a given genetic defect. As with vaccine viruses, this presumably will leave the germ-cell DNA unaltered, and therefore does not attack the defective gene as such. If we can cope with the disease, should we bother about the gene? Or may we not leave that problem to another generation.

There has been much to-do about another theoretical possibility, "cloning" a man, as might be done by the renucleation of a fertilized egg with

76

a somatic cell nucleus from an existing individual. Similar experiments have been successfully completed with frogs, and are being attempted with mice. Such experiments with laboratory animals will surely be very fruitful if the technique can be developed. It would also have enormous value in livestock breeding, just as cloning (propagation by cuttings) is a mainstay of horticulture. Until such experiments have been pursued in some depth with other animals, it is merely a speculative game to discuss applying such reproductive novelties to man. There is no urgent social problem to be addressed by such a technique. It does serve as a metaphor to indicate that future generations will have infinitely more powerful ways than we do to deal with whatever they may regard as socially urgent issues of human nature. We can therefore focus more confidently on dealing with the distress of individual human beings in the immediate generation. The metaphor also suggests that intrusive genetic engineering, if it is pursued for any other reason, will have plenty of policy problems to digest even before the "technology" has reached the point of detailed synthesis of genotypes by design.

Finally, medical scientists in general are fully aware and have fully assimilated ethical concerns about the application of new techniques in man, by contrast to experimental animals. For a long time, it has been known that one could operate on the brain in such "interesting" ways as dividing the corpus callosum with the possibility of the development of autonomous "intellects" in the two hemispheres. It would be unthinkable to apply such surgical technology to man without the persuasion and conviction that it would be for the benefit of the patient-subject. We will not be given the benefit of the doubt in public discussions of such questions; there are many influential people who really believe that "anything feasible will be done," and we may have to restate the obvious many times in reviewing the ethical constraints on possible experimentation.

To return to the "clone-a-man" metaphor: in my view, we simply do not know enough about the question at either a technical or an ethical level (and these are intertwined) to dogmatize about whether or not it should ever be done. Certainly it cannot be thought of within the framework of our generally accepted standards of medical ethics, unless (1) we can make and communicate a reasonably confident prediction of the outcome, and, more important (2) it has the informed consent and serves a reasonable humanitarian purpose for the individuals who are involved. In genetic matters, this must include the interests of the prospective newborn as well as of his parents and of the community. If we demand that he be represented in person, then no one could reasonably be allowed to be born whether by "natural" sexual fertilization by the design of his parents or otherwise. Cloning-a-man is one of the least important questions I can think of; "who must be held accountable for the next generation and how" may be the most important question.

Human Evolution by Voluntary Choice of Germ Plasm

For some decades the term *eugenics* has been in such disrepute, as a result of its spurious use in support of the atrocities committed by those with class and race prejudices, that few responsible students of evolution or genetics have dared to contaminate themselves by mentioning it, much less by dealing with the subject except in condemnation. However, it is now high time to take new stock of the situation. For the odious perversions of the subject should not blind us longer to a set of hard truths, and of genuine ethical values concerning human evolution, that cannot be permanently ignored or denied without ultimate disaster. On the other hand, if these truths are duly recognized and given expression in suitable policies, they may open the way to an immeasurable extension and enhancement of the potentialities of human existence.

In view of the signal defeat in World War II of the leading exponents of racism—a defeat which is still gathering momentum—and the declining prestige afforded in the Western world to the claims of aristocratic or bourgeois class differentiations, it at last becomes feasible to return, in a more reasonable spirit, to the theme of prospective human biological evolution. Moreover, for this job of re-examination we are now provided not only with a better understanding of genetic and evolutionary principles but also with a considerably reformed structure in most Western societies, liberalized mores, a heightened freedom of discussion, and a marked improvement in technologies, all of which combine to make possible approaches that earlier would have seemed out of the question.

It was Darwin who pointed out that modern culture is causing a relaxation and perhaps even a reversal of selection for socially desirable traits, and he expressed himself rather pessimistically about the matter, although in his time this process must have been much less pronounced than it is nowadays. His cousin Galton, impressed by Darwin's arguments concerning evolution in general as well as by those per-

SCIENCE, 1961, Vol. 134, pp. 643-649. Copyright 1961 by the American Association for the Advancement of Science.

taining to man, but unwilling to accept defeat or frustration for humanity on this score, proposed the idea that the trend might be counteracted consciously. For this course of action he coined the term *eugenics*, included within which he understood all measures calculated to affect the hereditary constitution in a favorable way. As he pointed out, these measures might be of very diverse kinds, lying not only in such fields as medicine but also in education, economics, public policy in general, and social customs, although he did not contemplate drastic changes from the mores of that Victorian age.

Unfortunately, although Galton realized to some extent the influence of the social and familial environment in the shaping of people's psychological traits, he was not sufficiently aware of the profundity of the environmental control. He therefore made the naive mistake, so widespread in his day, of looking upon the performances of different ethnic, national and social groups as indicative of their genetic capabilities and inclinations, although there were plenty of object lessons of the comparatively rapid transference of cultures that should have taught him better. Later, it was the madness of such out-and-out racists and so-called "social Darwinists" as Madison Grant, Lothrop Stoddard, Eugen Fischer, Lenz (*1*), and the Hitlerites which, carrying these prejudices much further, brought such odium upon the whole concept of eugenics as to run it into the ground.

Meanwhile, a large group of psychologists, represented by the Watson school, and of other social scientists, social reformers, socialists, and communists, all of them persons of egalitarian sympathies, impressed by the enormous potency of educational and other cultural influences, and regarding all eugenics as a dangerous kind of reaction that threatened their own roads to progress, popularized the idea that differences in human faculties are of negligible consequence not only as between different peoples and social classes but even as between individuals of the same group. They held that genetics in man could be allowed to take care of itself. And even where some genetic defects were admitted to exist, it was maintained that improved medical, psychological, and other cultural ministrations would provide sufficient remedies for them. Moreover, added the many Lamarckians among these groups, the improvements thereby acquired would eventually pass into the hereditary constitution. In this way, not only would all men become equalized but they would rise to ever higher biological as well as cultural levels. By about 1936 it had become a dire heresy among the official communists to dispute this line of argument, and the word *eugenics* had become a favorite symbol of all that is vile.

It is no wonder that earnest students of evolution and genetics, confronted by the mighty currents of these contending movements, which had advanced into the area of power politics, seeing how intertwined truth and error had become, and aware that their own views would almost certainly be misconstrued, tended to withdraw into their ivory towers and to refuse to discuss seriously the possible applications of genetics to man. It is to their credit, however, that only a few floated with the current that happened to be around them. But even fewer tried to contend with that current by raising the voice of reason, for that way lay the path to martyrdom.

Perhaps the last attempt made, up

until the past few months, to present an appraisal of eugenics undistorted by extremist politics was the drafting of the "Geneticists' Manifesto" (2) of 1939, signed by about a score of participants at the International Genetics Congress held in Edinburgh just as the curtain began to rise on World War II. In this document it was pointed out that by far the greatest causes of differences between human groups in regard to psychological traits were environmental, predominantly cultural, whereas in the causation of such differences between individuals within the same group, both environmental and genetic factors were very powerful, and often comparable in their potency of action. The need for far-reaching reforms—for affording more nearly equal opportunities to all groups as well as to all individuals and for removing biases—was stressed in this document, not only for the sake of persons directly concerned but also to provide a groundwork for the truer assessment of genetic differences, in the interest of more soundly based eugenics.

These reforms in society were also needed, it was pointed out, for the attainment by the population of a sounder set of values, applicable equally to eugenic and to cultural purposes: values by which active service and creativity would be regarded more highly than either passive submissiveness or self-aggrandizement. The "Geneticists' Manifesto," far from discarding the concept of eugenics per se, acknowledged that it afforded, when rightly used, a means of making far-reaching human progress of a kind that must complement purely cultural advancement. Beyond that, genetic improvement was even affirmed to be a right which future generations would consider those of the past who were

aware of the situation as having been obligated to accord them, just as they in turn would consider themselves as being similarly obligated to their own successors. This obligation would not be regarded as a burden, however, but rather as a high privilege and a challenge to their creativity. At the same time, it was recognized that adequate implementation of eugenic policies also required a clearing away of the ancient heritage of superstition and taboos that hitherto had so obstinately enshackled human usages and preconceptions in matters of sex and reproduction.

It is true that our own world of today is still grievously beset with the old inequalities, prejudices, and mummeries. However, all peoples have by now seen the handwriting on the wall that spells the end of these irrationalities. For modern technologies have, on the one hand, made it too dangerous for the world to remain divided. They have, on the other hand, provided the means for achieving an unparalleled interdiffusion of techniques, ideas, personnel, education, and socioeconomic organization, and for raising standards of living. In the process, provincialisms are at last being ground down, though not without much friction. Opportunities, educational, economic, and social, are being extended ever more effectively to the more depressed social classes and ethnic groups in our own country and elsewhere. A real effort is being made to bring the viewpoint of science home to the general population. The battles that superstition is still winning take place ever closer and closer to its heartland, as was so well depicted in the moving picture on the Scopes trial, *Inherit the Wind*. And it is even becoming permissible to debate seriously matters that in the days before nuclear

80

weapons, space ships, Kinsey, and the Darwin centenary were taboo among all nice people.

Thus the scene has at last been shifted to such an extent as to make it fitting to re-examine even such a scandalous subject as eugenics, with a view to preparing the new forces now arising in the world to deal with it both realistically and humanistically. For, as we shall see, cultural progress of the kinds mentioned has already proceeded far enough here and there to make the beginnings of a new approach to the subject possible, and the attitudes now formed and the preparations now made may presently lead, when the time is riper, to more salutary developments in this field than could ever before have occurred.

Contradictions in the

Traditional Eugenic Methods

Let us first examine the methods by which it has hitherto been thought that eugenics might operate. These methods have taken their cue from the natural selection of the past. All evolution has had its direction determined in some way by the force of selection. Selection chooses among the materials available to it, namely, diverse mutations, which occur in a manner that is fortuitous so far as their adaptation to the needs of the organism in the given situation is concerned. Selection acts entirely through differential multiplication, but this process can be conceptually divided into two parts, namely, unequal rates of survival (or, conversely stated, of mortality), on the one hand, and unequal rates of reproduction of the survivors, that is, differential fertility (or, conversely stated, differential infertility), on the other hand.

Eugenists have therefore distinguished between two conceivable methods—differential control over mortality, and differential control over reproductive rate. The first method, however, although practiced by the Spartans and by primitive tribes who destroyed infants regarded as undesirable, is universally acknowledged to be inconsistent with the respect for human beings that forms an essential part of civilization. It might be contended that artificial abortion is an intermediate method, but everyone recognizes this also to be an undesirable means where other procedures are available. Essentially, then, this has left for eugenics the second alternative: that of a qualitatively differential control over reproduction prior to or at conception.

In Galton's time, before the advent of modern contraceptive techniques, it was indeed rather visionary to conceive of people's reproduction being governed in the interests of the progeny. For this could be done only by the drastic method of surgical sterilization, of a type that interfered with the sexual life, or by such consummate self-control as voluntary abstention from intercourse, or from its completion.

The development of techniques for cutting or ligating the tubes that conduct the mature reproductive cells afforded less objectionable means of sterilization, but this procedure was still usually regarded by most people —rightly or wrongly—as too irrevocable, except, perhaps, for persons who were mentally or morally hopelessly irresponsible. For them, enforced operations of this type were legalized in some regions, although it was rightly pointed out that there was grave danger of abuse of the practice unless it were confined to the most extreme cases. For attitudes that seem wrong in one place or setting may seem right elsewhere, and nonconformists may at times have moral standards superior, in a longer perspective, to those of the majority who condemn them. Thus, the amount of sterilization resulting from legal applications of an advisable kind would be so minute as to have very little eugenic influence.

However, the invention of fairly practicable artificial means of voluntary contraception opened up much wider possibilities for the control of reproduction in economically developed countries. As we all know, advantage has been taken of these techniques on a large scale, and they have become one of the indispensable procedures whereby the general standard of living has been so greatly raised. Still more practicable means of contraception seem at last to be on the way, thanks to the efforts of a handful of devoted scientists, and they cannot come too soon, for it is imperative to make similar benefits possible in the less developed regions.

But although contraception that is used for the enhancement of cultural benefits through the control of population quantity is at the same time a *potential* instrument for the improvement of genetic quality, such improvement does not occur unless the contraception is specifically aimed in this direction. To be sure, this purpose might be achieved if the individual couples concerned were to reach their decisions about how many children to have in a highly idealistic spirit, one guided by almost heroic self-criticism and wisdom. We shall presently consider whether or not it is realistic to expect this. A second proposal has been that of altering the economic and social system in such a way that people of higher gifts and greater natural warmth of fellow feeling—that is, the genetical-ly more highly endowed—would be normally led into occupations and modes of life more conducive to having a large family. Conversely, the organization of society which this view would hold to be ideal would tend to lead persons less well endowed to choose, of their own accord, situations in life that would encourage them to expend their energies in other pursuits than reproduction, and would give them less inducement for raising families.

These two approaches, the individual and the societal, are, of course, not mutually exclusive, and most 20th-century eugenists have advocated a combination of the two. But let us examine each of them more closely. First, as regards the individual approach, which is supposedly to be adopted by people in general once they have been well educated in matters of evolution and genetics, it should be acknowledged that people in general can in fact be taught to take pride in making great sacrifices for what they recognize to be a great cause, especially when they win social approval thereby. This has often happened in times of war as well as after social revolution. However, it seems asking almost too much to expect those individuals who are *really* less well equipped than the average, in mentality or disposition, to acknowledge to themselves that they are genetically inferior to their neighbors in these respects, and then to publicly admit this low appraisal of themselves by raising no family at all or a smaller one than normal, especially since at the same time they would often be thwarting a natural urge to achieve the deep fulfillments, accorded to their neighbors, that go with having little ones to care for and bring up. Moreover, those with physical impairments would likewise tend to rationalize the situation, by thinking that they possessed some superior psychological qualities that more than compensated for their physical defects.

In fact, then, the ones most likely to comply with the idea of restraining their own reproduction would be those who had such strong social feelings, such a sense of duty, so high a standard of what is good, so little egotism, and such an urge for objectivity, as actually to lean over backward and so underrate themselves. Thus we would be likely to lose for the next generation much of what might have been its best material.

On the other hand, for many of the really gifted there are often unusual opportunities for achievement, for rich experiences, and for service along other lines than those of bringing up a large family. Hence it would be only human of them, even though they were in sympathy with eugenic ideals, to expend a larger share of their energies in these other ways than does the average man or woman, for whom the home is often both a refuge and the chief stage on which to express leadership. In view of these considerations, it is not at all surprising that eugenic practices of this intentional, personal type, that require a correlation between the size of one's family and one's realistically made appraisal of one's genetic endowments, have made so little headway, even where they were approved theoretically. Thus, even among eugenists themselves, one seldom finds much evidence that these principles are being acted upon.

What, then, about the proposal that our society should introduce features into its structure whereby the more gifted, the abler, and the more socially minded would find conditions more conducive to their raising a large family, while those less capable or relatively antisocial would tend automatically to be deflected from family life? Surely we would not want a dictatorship to institute such a system, for dictators are oftener wrong than right in their decisions. Moreover, their subjects are not able to become truly men, in the all-round sense, and those who shine under such circumstances are not likely to be the wise and the responsible.

Under a democracy, on the other hand, is it not likely that "the common man" will refuse to subject himself to such manipulations? Certainly if the proposal took some such crude form as a subsidy for the raising of children, allotted to those who already occupied better positions or who had scored higher on certain tests, it would rightly be resented and defeated as discriminatory by the great majority. And even if subtler forms of influence were used, such as special aids to family life for those in occupations requiring greater skills, responsibility, or sacrifice, there would soon be a clamor on all sides to have these advantages extended to every responsible citizen. No, we can hardly use democracy to support any kind of aristocracy. To be sure, ways might eventually be found to reduce the present strong negative correlation between educational or social achievement and size of family, and these would be all to the good. But no major formula is in sight for restoring the greater family size of the fitter, while retaining that most essential feature of our culture—the extension to all of mutual aid based on the most advanced technologies available.

It might seem to follow that we have now, as a result of our improved techniques for living, reached an inescapable genetic cul-de-sac. It might be

83

concluded that we should therefore confine ourselves entirely to the immediate job on hand—the pressing and rewarding one of all social reformers and educators — that of making the best of human nature as it is, the while allowing it to slide genetically downhill, at an almost imperceptible pace in terms of our mortal time scale, hoping trustfully for some miracle in the future.

The New Approach: Germ-Cell Choice

However, it is man who has made the greatest miracles of any species, and he has overcome difficulties arising from his technologies by means of still better basic science, issuing in still better technologies. And so in the case of the genetic cul-de-sac of the present day, he has even now possessed himself of the means of breaking through it. For he is no longer limited, like species of the past which had the family system, to the two original methods of genetic selection applying to them: that of differential death rate on the one hand, and differential birth rate or family size, on the other hand. He has now given himself, in addition, the possibility of exerting conscious selection by making his own choice of the source of the germ cells from which the children of his family are to be derived (3–5). At present, this choice is confined to the male germ cells, but there are indications that with a comparatively small amount of research it might in some degree be extended to those of the female as well.

It is pretty common knowledge nowadays that some tens of thousands of babies have already been born, in the United States alone, that were derived by "AID," that is, by artificial insemination with sperm obtained by the physician from a donor chosen by him, but whose identity was kept unknown to all others, including even the parents. In the great majority of these cases the husband had been found to be irremediably sterile. And although, in view of the prevalence of the traditional mores, the whole matter was kept secret from the children themselves, both members of the couple had in these cases been eager to avail themselves of this opportunity to have one or more children. This method has, of course, been allowed only to those likely to make good parents—or, shall we say, good "love parents," as distinguished from "gene parents"? Moreover, follow-up studies have shown that these parents did truly love their children, as the children did their parents. It is noteworthy that this proved to be as much the case for the father as for the mother, and that the marriage was strengthened thereby.

Here, then, we see repeated what is typically found in those cases of early adoption in which the children have been genuinely desired. However, this "pre-adoption" (as Julian Huxley has termed it) is likely to prove even more binding and satisfying than "post-adoption." And the method of pre-adoptional choice, despite its relative crudity at the present time, has demonstrated its capability of producing a superior lot of children. Thus, the couples that practice it have made a virtue of necessity by inducing the genesis of children of whom they can usually be even prouder than of the children they would have had if they had been free from reproductive infirmity.

Recently, as more people have become alive to matters of genetics, an increasing number of couples have resorted to AID when the husband, although not sterile, was afflicted with some probably genetic impairment, or likely to carry such an impairment that had been found in his immediate family. Similarly, those with incompatabilities in blood antigens have also made use of the method with good results. In these ways a beginning has already been made in the conscious selection of germinal material for the benefit of the progeny. It is to be expected that many more people will seek such advantages for their prospective children when there are available for creating them the germ cells of persons who are decidedly superior in endowment to the normally healthy and capable persons that are commonly sought as donors by the physicians of today. For there is no physical, legal, or moral reason why the sources of the germ cells used should not represent the germinal capital of the most truly outstanding and eminently worthy personalities known, those who have demonstrated exceptional endowments of the very types most highly regarded by the couple concerned, and whose relatives also have tended to show these traits to a higher-than-average degree. How happy and proud many couples would be to have in their own family, to love and bring up as their own, children with such built-in promise.

Today, of course, most people have such an egotistical, individualistic feeling of special proprietorship and prerogative attaching to the thought of their own genetic material as to be offended at the suggestion that they might engage in such a procedure. No one proposes that they do so as long as they feel this way about the matter. However, they should not try to pre-vent others who would welcome such an opportunity from ordering their lives in accord with their own ideals. And as the prejudice against the practice gradually dwindles, the manifest value of the results for those who had participated in it would appeal to an increasingly large portion of the population.

In this connection it is important to bear in mind that there is no such thing as a paternal instinct in the sense of an inherent pride in one's own genetic material or stirps. Some primitive peoples, even including a few still in existence in widely separate regions, have had, strange as it may seem to us, no knowledge that the male plays a role in the production of the child, much less have they had any conception of genetic material or genes. Thus, among some of them the mother's brother has effectively filled the role of father in regularly caring for mother and children. Moreover, among some peoples, such as the Hawaiians, children are rather freely adopted at an early age into other families, into whose bosom they are warmly, fully, and unambiguously accepted as equals in every way to the natural children. It is "second nature," but not "first nature," for us in our society to exalt our own stirps.

It is, however, "first nature" for men and women to be fond of children and to want to care for them, and more especially, those children with whom they have become closely associated and who are dependent on them. If the love of a man for his dog, and vice versa, can go to such happy lengths as it often does, how much stronger does the bond normally become between the older and younger generations of human beings who live together. And since, in the past, the

85

children *have* usually been those of the parents' own stirps, it has been a natural mistake to suppose that these stirps, rather than the human associations of daily life, formed the chief basis of the psychological bonds that existed between parent and child. Yet, as our illustrations have shown, this view is incorrect, and a family life of deep fulfillment can just as well develop where it is realized that the genetic connection lies only in our common humanity.

The wider adoption of the method of having children of chosen genetic material rather than of the genetic material fate has chanced to confer on the parents themselves implies, of course, that material from outstanding sources become available by having it stored in suitable banks (*3, 6, 7*). It would be preferable to have it in that glycerinized, deep-frozen condition developed by modern technology, in which it remains unchanged for an unlimited period without deterioration. It is true that research is badly needed for finding methods by which immature germinal tissue can, after the deep-freezing which it is known to survive, be restored to a state where it will multiply in vitro and subsequently produce an unlimited number of mature reproductive cells. Even without this further development, however, the way is already open, so far as purely biological considerations are concerned, for gathering and inexpensively storing copious reserves of that most precious of all treasures: the germinal material that has formed the biological basis of those human values that we hold in highest regard.

The high potential service to humanity represented by the pre-adoption of children should carry with it the privilege, for the parents, of having a major voice in choosing from what source their adopted material is to be derived. Surely, if they have ever had the right to produce, willy-nilly, the children that would fall to their lot as a result of natural circumstances, they should have the right of choice where they elect to depart from that haphazard method. They certainly would not wish knowingly to propagate manifest defectives, and, being idealistic enough to undertake this service at all, they would in most cases be glad to give serious consideration to the best available assessments of the genetic probabilities involved, as well as be open to advice regarding relative values and needs.

It would be made clear to parents that there is always an enormous amount of uncertainty concerning the outcome in the genetics of an organism so crossbreeding as man, especially since the most important traits of man are so greatly influenced by his cultural environment. Nevertheless, facing this, they would realize that the degree of promise was in any such case far greater than for those who followed the traditional course. It would be in full awareness of this situation that they would exercise their privilege of casting the *loaded* dice of their own choosing.

This kind of choice means that the physician can no longer be the sole arbiter of destiny in this matter. Clearly, if the couple are to accept their share of the responsibility and privilege here involved, the practice of keeping the donor unknown to them must be relinquished in these cases. Knowledge of the child's genetic lineage will also be needed later, so that sounder judgments may be reached concerning his genetic potentialities in the production of the generation to follow his own. This lifting of the veil of secrecy will become ever more practicable, and in fact even necessary, as the having of children by chosen genetic material becomes more widely accepted and therefore more frequent. Moreover, the attitude of others toward the couples who have employed this means of having their children will gradually become one of increasing acceptance and then of approbation and even honor.

Today's fear that knowledge by the mother of the identity of the gene father may lead to personal involvement between the two, to the detriment of normal family life, will recede when the gene source is remote in space or time, as when the germinal material has been kept in the deep-frozen state for decades. This procedure will also allow both the individual worth of those being considered as donors, and their latent genetic potentials, to be viewed in better perspective, and will reduce the danger that choices will be based on hasty judgments, swayed by the fads and fashions of the moment.

Let us see in what ways this method of reproduction from chosen material tends to avoid the difficulties that are encountered in attempting to reconcile traditional reproduction with the interests of genetic quality by somehow controlling the size of families. For one thing, as previously pointed out, most men would resist accepting and acting on the conclusion that they are below their next-door neighbor, or below average, in genetic quality. They would particularly resist the idea that they themselves are in that lowest fifth which would be required to refrain from having children if an equilibrium of genetic quality were to be maintained in the face of a 20-percent mutation rate. Yet most of these same people would willingly accept without resentment the idea that they are not among the truly exceptional who conform most closely to their own ideals. And so, when encouraged by the community mores, they would be glad and proud to have at least one of their children derived, by choices of their own, from among such sources. Thus they would continue to have families of a size more nearly conforming with their inclinations.

On the other hand, the worthy but humble, those who might otherwise, from overconscientiousness, limit their families unduly, would often be eager to serve as love parents. And although in that capacity they would tend to derive the germ cells from outside sources, they would be especially likely to have a well developed sense of values and so to choose sources even worthier than themselves. Finally, the really highly endowed but realistic would not be confronted with the sore dilemma of choosing between exercising their special gifts, on the one hand, or having the large family to which genetic duty seemed to obligate them, on the other hand. For their germinal material would tend to be sought by others, if not in their own generation, then later, and to a degree more or less in proportion to their achievements. Thus they would be freed to give their best services in whatever directions they elected.

In all these ways, the diverse obstacles encountered by a eugenics that tried to function by means of a consciously differential birth rate—that is, by adjustment of family size—would be avoided. Thereby, a salutary separation would be effected between three functions that often have conflicting needs today. These are, first, the choice of a conjugal partner; this should be determined primarily by sexual love, companionability, and compatible mentality and interests. Second, there is the determination of the size of the family; this should depend largely on the degree of parental love that the partners have, and on how successfully they can express it. Third, there is the promotion of genetic quality, both in general and in given particulars; these qualities are often very little connected with the first two kinds of specifications. By thus freeing these three major functions from each other, all of them can be far better fulfilled. Under these circumstances the conjugal partners need not be chosen by criteria in which a compromise is sought between the natural feelings and considerations of eugenics. Neither need the family size be restricted or expanded according to eugenic forebodings or feelings of duty. Yet at the same time there can be a far more effective differential multiplication of worthy genetic material than in any other humanly feasible way.

Further Prospects

It is likely that the avoidance of the effects of sterility will not be the only door through which such a change of mores will be approached. Facilities for keeping germ cells, suitably stored below ground in a deep-frozen condition, in areas relatively free from radiation and chemical mutagens, may well be provided in our generation for an increasing number of people (8). Among these would be persons subject to the growing radiation hazards of industry, commerce, war, and space flight, and those exposed to the as yet unassessed hazards of the chemical mutagens of modern life. The same means would greatly retard the accumulation of spontaneous mutations which probably occurs during ordinary aging. Thus, wives may in time demand such facilities for the storage of their husbands' sperm. These facilities would be provided not only for the sake of reducing mutational damage but also as a kind of insurance in the event of the husband's death or sterility. In these ways great banks of germinal material would eventually become available. They would be increasingly used not only as originally intended but also for purposes of conscious choice. Moreover, some of these stocks might become recognized as especially worthy only after those who had supplied them had passed away.

The cost of storage is, relatively, so small that failure to make such a provision will eventually be considered gross negligence. As Calvin Kline (7) points out, this will be especially the case where (as in India today) voluntary vasectomy becomes more prevalent as the surest and, in the end, the cheapest means of birth control. For when vasectomy is complemented by stores of sperm kept in vitro, the process of procreation thereby achieves its highest degree of control—control not subject to the impulses of the moment but only to more considered decisions.

It is true that most people's values, in any existing society, are not yet well enough developed for them to be trusted to make wise decisions of the kind needed for raising themselves by their bootstraps, as it were (9). But this type of genetic therapy, of "eutelegenesis," as Brewer termed it when he advocated it in 1935 (4), is certainly not going to spring into existence full fledged overnight. It will first be taken up by tiny groups of the most idealistic, humanistic, and at the same time realistic persons, who will tend to have especially well developed values. This mode of origination of such practices was pointed out by Weinstein in "Palamedes" in 1932 (10). These groups will tend to emphasize the most basic values that are distinctive of man, those that have raised him so far already, but which still may be enormously enhanced. Foremost among these are depth and scope of intelligence, curiosity, genuineness and warmth of fellow feeling, the feeling of oneness with others, joy in life and in achievement, keenness of appreciation, facility in expression, and creativity.

Those who follow these lodestars will blaze the trail, and others will follow and widen this trail as the results achieved provide the test of the correctness with which its direction was chosen. Meanwhile, the world in general, through its reorganizations of society and education, is moving in the same direction, by a de-emphasis of its provincialisms and a consequent recognition of the supreme worth of these basic human values. For after all, these values have been prominent in the major ethical systems of the whole world.

At the same time, plenty of diversity will inevitably be developed. For each especially interested group will naturally seek to enhance its particular proclivities, and this is all to the good. But on the whole, the major gifts of man have been found to be not antagonistic but correlated. Thus we may look forward to their eventual union with one another in a higher synthesis. And from each such synthesis in turn, divergent branches will always be budding out, to merge once more on ever higher levels.

It may be objected that we have next to no knowledge of the genes for those traits we value most, and that their effects are inextricably interwoven with those of environment. As was acknowledged earlier, this is quite true. However, it has also been true in all the natural selection of the past and in the great bulk of artificial selection. Yet these empirical procedures, based entirely on the accomplishment of the individuals concerned, did work amazingly well. We can do a good deal better by also taking advantage of the evidence from relatives and progeny. Yet that evidence also is furnished mainly by accomplishments or output. Where those were high, the environment was, to be sure, usually favorable, but so was the heredity. And as, in our human culture, social reform proceeds and opportunities become better distributed, our genetic judgments will become ever less obscured by environmental biases, while at the same time our knowledge of genes will improve.

Meanwhile, the efforts of educators and the lessons of world affairs will serve to emphasize the same values for us. And these attitudes we will take over for our genetic judgments also. In this connection, another consideration deserves mention here. The preference which most parents will inevitably have for the genes of persons of truly remarkable achievement and character, rather than for those of the merely eminent or powerful, will at the same time serve to direct the stream of genetic progress toward the factors underlying creativity, initiative, originality, and independence of thought, on the one hand, and toward genuineness of human relations and affections, on the other hand. Otherwise the genetic movement might, as so often happens in other affairs of men, become directed toward skill at conformity, showmanship, and the dignified hypocrisy that often brings mundane success and high position (9, 10). This would have been a far greater danger in the case of the old-style eugenics.

But, it may be objected, does all this really represent conscious control in an over-all sense? Is it not merely a type of floating along in a chaotic manner, each straw making its own little movement independently of the rest, without a general plan or goal or stream? The answer is that humanity is as yet too limited in knowledge and imagination, too undeveloped in values, to see more than about one step ahead at a time. That step, however, can be discerned clearly enough, and by enough people, to give rise to a general trend in a salutary direction. And at the higher level to which each step taken will bring us we will be able to see an increasing measure of advance ahead. So we humans will achieve, not through dictation but through better general understanding and ever more clearly seen values, increasing mutual consent both concerning the means to be used and the aims toward which to orient. Thus an ever wider over-all view will emerge, and a surer, greater over-all plan, or

rather, series of plans. To create them and to put them into effect will then enlist our willing efforts. And the very enjoyment of their fruits will bring us further forward in our great common endeavor: that of consciously controlling human evolution in the deeper interests of man himself (11).

References and Notes

1. H. J. Muller, *Birth Control Rev.* **17**, 19 (1933).
2. *J. Heredity* **30**, 371 (1940); *Nature* **144**, 521 (1939).
3. H. J. Muller, *Out of the Night* (Vanguard, New York, 1935).
4. H. Brewer, *Eugenics Rev.* **27**, 121 (1935); *Lancet* **1**, 265 (1939).
5. H. J. Muller, *Perspectives in Biol. and Med.* **3**, 1 (1959); ———, in *Evolution after Darwin*, Sol Tax, Ed. (Univ. of Chicago Press, Chicago, 1960), vol. 2, p. 423.
6. H. Hoagland, *Sci. Monthly* **56**, 56 (1943).
7. C. W. Kline, unpublished address before the Society for Scientific Study of Sex, New York (1960).
8. H. J. Muller, in *The Great Issues of Conscience in Modern Medicine* (Dartmouth Medical School, Hanover, N.H., 1960), p. 16.
9. ———, *Sci. Monthly* **37**, 40 (1933).
10. A. Weinstein, *Am. Naturalist* **67**, 222 (1933).
11. This article was originally prepared as a lecture to be given on the invitational program of the Society for the Study of Evolution at its annual meeting in New York, 29 Dec. 1960, but adventitious circumstances prevented its delivery on that occasion.

BIOCHEMICAL AND CHROMOSOMAL DEFECTS YIELD TO PRENATAL DIAGNOSIS

by Paul T. Libassi

In the past, the best information that a clinician could provide to prospective parents was a prediction of whether or not their child was likely to be normal. In essence, odds were quoted. Today some 24 inherited biochemical defects as well as all of the major chromosomal abnormalities can be diagnosed prenatally with near certainty.

Amniocentesis—the extraction of fluid which surrounds the fetus—is the keystone of prenatal diagnosis, providing the raw materials from which diagnosis proceeds. Fetuses shed some cells into the amniotic fluid in which they are bathed. These cells can be retrieved as early as the eleventh week of fetal life and grown in culture. Subsequently, biochemical assays for enzyme deficiencies can be performed and chromosomes can be visualized to permit diagnosis of the various chromosomal aberrations.

Timing in dispute. The optimal time during pregnancy when diagnostic amniocentesis is most safely performed remains a matter of dispute. A compromise must be reached between the conflicting facts that the later an amniocentesis is performed, the safer it is, and the earlier that abortion occurs, the safer it is.

Dr. Fritz Fuchs, Chairman of the Department of Obstetrics and Gynecology at Cornell University Medical College maintains that during the 15th week of pregnancy sufficient amounts of amniotic fluid are available to permit the removal of 10-20 ml of fluid without causing adverse effects in either fetus or mother. Amniocentesis performed during the 15th week of gestation assumes that abortion should be performed no later than the 20th week and allows four to five weeks for cells to grow in culture and for diagnosis to be made. If biochemical or chromosomal determinations can be completed in less than four weeks time, amniocentesis should then be delayed for several weeks until the 16th or even 17th week of pregnancy.

Cells harvested from amniotic fluid are cultured for analysis—either biochemical or chromosomal—according to the following general scheme: Fetal cells suspended in the amniotic fluid are separated through centrifugation and, after all liquid has been drained, are placed in a culture medium. In the case of biochemical analysis, cells are grown in this culture medium until sufficient amounts are available for enzyme deficiency tests. Cells are harvested for chromosomal analysis when sufficient numbers reach the metaphase stage of division—the point at which each chromosome has divided into two mirror-

LABORATORY MANAGEMENT, 1972, Vol. 10, No. 9, pp. 20-24.

91

image halves but are still connected at the centromere.

The cells are run through a centrifuge again to separate them from the culture medium and are treated with colchicine to halt mitosis. A dilute salt solution is then added to the cell preparations to induce swelling and to cause the chromosomes to migrate apart. At this point the cells are fixed and stained and a chromosome spread is photographed through the microscope.

Prenatal diagnosis of biochemical defects represents, according to Dr. John W. Littlefield, Chief of the Genetics Unit at Massachusetts General Hospital, an unusual situation in medicine. The diagnostic information which provides the basis for any therapeutic measures to be taken is based upon a single laboratory test for an enzyme deficiency. Because of this, laboratories must take inordinate precautions to develop successful cultures and enlist the aid of confirmatory procedures whenever possible. The preparation of successful cultures depends in great part on the amount of amniotic fluid from which cells are havested. Opinions on just how much fluid should be drawn vary.

Dr. M. Neil Macintyre's laboratory at Case Western Reserve University obtains as much as 40 cc of fluid, a quantity large enough to maintain cultures in as many as six or seven flasks. According to Dr. Macintyre, several important advantages result from multiple culture preparations. The possibility of an erroneous diagnosis due to maternal cell contamination is greatly minimized, and the likelihood of an earlier diagnosis is enhanced by the growth of a greater volume of cells.

Culture methods vary from laboratory to laboratory. A choice of culture techniques should ultimately depend upon those methods which supply the most accurate data in the shortest pos-

sible time, according to Dr. Macintyre. The cytogenetic laboratory is responsible for production of cultures to be utilized for biochemical analysis and for chromosomal preparations of sufficient volume and quality to facilitate a speedy and accurate diagnostic evaluation. Dr. Littlefield's laboratory has found that a medium supplemented with 15 percent calf serum offers the best results. Cultures at his Massachusetts laboratory are re-fed twice a week, a procedure which appears to enhance the growth of even single cells.

Problems. Several types of cells from epithelial-like to fibroblast-like are present in amniotic fluid. Problems arise because clinicians cannot be certain that in the case of female fetuses the cells that have been cultured are fetal and not maternal. Researchers have suggested that several developments would alleviate this problem and confirm any discrepancies.

Specific and clear-cut differentiation of the isozyme content of fetal and adult cells would be of important practical value. Studies have documented differences in the activity of enzymes between cultured amniotic cells and adult skin fibroblasts. In addition, Dr. Littlefield suspects that different isozyme patterns, in addition to quantitative changes in enzyme activity, are displayed by amniotic cells. If an isozyme pattern peculiar to fetal cells could be defined, it would provide an important confirmatory adjunct. Dr. Littlefield adds that such an isozyme pattern should be detectable in a spread of relatively few cells so that confirmation could be made in time to permit drawing a second specimen of amniotic fluid if necessary.

An alternative to specific isozyme patterns to confirm fetal cells is visualization of the fetus before or during amniocentesis, thereby guaranteeing that fetal cells have in fact been obtained. Several scientists are molding fiber glass optics into an instrument called a "fetal amnioscope" just for such purposes.

Dr. Henry Nadler, Chairman of the Department of Pediatrics at Northwestern University and a pioneer in amniocentesis, reports that bioengineers at Northwestern have developed an instrument that can be inserted into a needle only slightly larger than the size presently used for amniocentesis. This instrument has facilitated visualization of an increased number of external morphologic anomalies and in the future may allow clinicians to actually remove a snip of fetal tissue assuring that cells being cultured are in fact fetal cells.

Refinements. In standard chromosomal analysis, the photomicrograph is enlarged and the chromosome images are cut out and arranged on a white card into 22 pairs of homologous chromosomes plus the two sex chromosomes. Pairs are arranged in sequential order according to morphological peculiarities based upon size, shape and ratio of arm length.

Several refinements of this manual procedure have since rendered the construction and examination of the karyotype both less time-consuming and more objective. Dr. Robert S. Ledley of the National Biomedical Research Foundation has developed a computer program which can analyze a photomicrograph with complete accuracy in less than 40 seconds.

The computer is programmed to count the total number of chromosomes in each cell and to measure and notate their morphological features. The computer then matches chromosome pairs according to these features and classifies each pair according to a standard karyotype sequence. In this manner, the computer is capable of evaluating small

but clinically significant chromosomal abnormalities which are often undiscernible by the human eye. Dr. Peter W. Neurath of New England Medical Center, Boston, also has developed a computerized chromosome analysis system (*Lab. Mgt.,* Sept. 1970).

Dr. Ledley is presently developing a special-purpose computer that will automate the entire procedure from selection to analysis. Such a system would eliminate any degree of subjectivity involved in the selection of cells to be photographed for analysis. A scanner linked directly to the microscope would automatically select and photograph the appropriate cells. The only problem with the system at this point seems to be the costs involved. Dr. Harold Nitowsky of the Albert Einstein Medical Center explains, "In terms of cost analysis, the manual procedures are certainly more feasible at this point."

The discovery of banding patterns unique for each individual pair of human chromosomes has resulted in a reliable and accurate method for identifying and aligning chromosomes on a karyotype. Dr. Margery Shaw at the University of Texas has found that chromosomes stained with a Giemsa stain display discrete and highly specific banding patterns in characteristic sections of the chromosome arm. Not only does this technique permit accurate homologous pairing, but banding patterns can reveal chromosomal defects impossible to discern by morphological characterization alone.

Microassays. One of the largest problems of prenatal diagnosis involves the relatively large volume of cells required for biochemical assay and the time required to grow them. As long as six weeks may be needed to culture sufficient amounts of cells for the types of biochemical assays currently available. The development of microassays for the

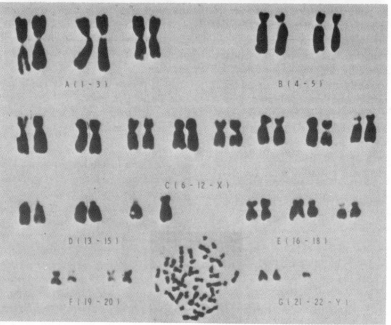

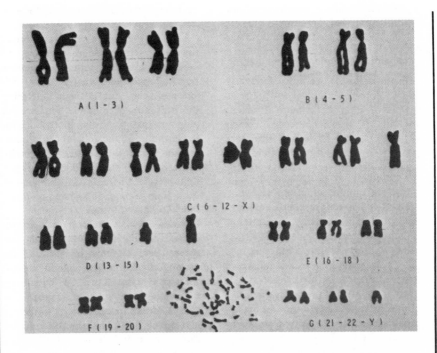

Dr. Henry Nadler confers with a husband and wife considering parenthood (top). The Northwestern University pediatrician explains that the mother is a translocation carrier whose children are likely to have Down's syndrome. Her chromosome karyotype (middle) indicates an abnormality in the No. 21 chromosome. A karyotype of the fetus (bottom) indicates that Down's syndrome will result from the extra No. 21 chromosome. (NIH photos.)

various enzymes involved would solve this problem to a great extent. Dr. Nadler is encouraged by the development of histochemical techniques sensitive enough to permit analysis of significantly fewer cells than are now required. If these techniques prove accurate, as few as five cells taking only three to four days to grow could be assayed.

Although analysis of uncultured amniotic cells is possible and has been performed with a degree of success, most work to date has been done with cultured cells. Researchers indicate that the reliability of a diagnosis on the basis of uncultured cells is questionable and should be avoided. Some cells harvested from the amniotic fluid are dead and can therefore lead to an erroneous diagnosis.

Uncultured cells are most frequently used to determine sex of the fetus as early as possible to facilitate the management of pregnancies in women heterozygous for X-linked disorders such as hemophilia. Amniotic fluid cells are harvested from cultures following the first day of incubation and are stained with fluorescent dyes for sex chromatin. This staining technique reveals the presence of a Y chromosome, indicating that the fetus is male. However, in cases where no fluorescence is present, conclusions based upon this technique alone are equivocal. The Y chromosome may have failed to fluoresce for a variety of reasons, the fetus may be female or only maternal cells may be present in the fluid. Furthermore, it is unlikely though possible that an X-linked disorder such as Turner's syndrome may interfere with a correct analysis of sex determination. For these reasons, researchers such as Dr. Macintyre recommend that sex chromatin analysis of uncultured cells be confirmed through karyotypes of cultured cells before diagnosis is finalized.

A problem area of particular concern to many clinicians results in cases of multiple pregnancies. In these situations there is a distinct possibility that amniotic fluid will be inadvertently sampled from only one of the amniotic sacs and the resultant cell cultures will there fore represent only one fetus. The development of ultrasonic techniques has recently made detection of multiple pregnancies possible early in gestation. As early as the eighth week of pregnancy ultrasonic probes can scan the surface of the abdomen and bounce back a series of echoes which can then be interpreted to identify and localize the fetuses.

Sound waves pass through abdominal tissue at variable speeds according to density. From the pattern of echoes bounced back and viewed on a television screen, doctors can determine the presence of multiple fetuses and pinpoint their positions in the uterus. Localization of the fetus affords information for a more effective amniocentesis and by alerting doctors to multiple pregnancies indicates the need to draw fluid from each sac and to maintain separate cultures.

Caution. Despite the fact that amniocentesis has come a long way in the last several years—two years ago an estimated 200-300 had been performed in the world; today some 1500 have been successfully carried out in the U.S. alone—most scientists and clinicians agree that the technique still remains rather crude and should not be undertaken by any except those with a reasonable amount of experience and expertise.

Dr. Nadler recommends that a series of requirements be satisfactorily fulfilled before intrauterine diagnostic measures are undertaken. These include an obstetrician experienced in the technique of amniocentesis, laboratory personnel experienced in cultivating amniotic fluid cells and in selecting and developing reliable spreads for karyotypes, technicians experienced in performing the specific diagnostic tests required, and the ability to provide suitable treatment. Treatment generally consists of therapeutic abortion. Dr. Henry Kirkman, Chief of Laboratories at North Carolina Memorial Hospital reports that amniocentesis is not performed on those patients for whom abortion is out of the question for either religious or personal reasons.

Diagnostic amniocenteses are recommended by physicians at the Genetics Referral Center of the University of California at San Diego on the basis of the following statistical portrait: a previous history of hereditary disorders; the mother is a known or suspected carrier of an X-linked disorder; the parents are known carriers of a chromosomal translocation; a previous child was born with a chromosome abnormality; the maternal age at conception was 35 or more years.

Not a total answer. Nor does amniocentesis provide the answers for all genetically related disorders. The two most common genetic defects, sickle cell anemia and cystic fibrosis, cannot as yet be detected prenatally with any degree of certainty. Methods for both of these are on the horizon and it seems only a matter of time before they are perfected.

Dr. Barbara Bowman at the University of Texas has demonstrated that serum from patients afflicted with cystic fibrosis inhibits the action of oyster cilia. The technique has not, however, yielded clear-cut separations between carriers and those afflicted with the disease. When these and other problems have been ironed out, Dr. Bowman hopes to adapt the technique to amniotic fluid for diagnosis of the disease *in utero.*

Prenatal diagnosis of sickle cell anemia presents different problems. According to Dr. Nitowsky, methods for identifying the sickle cell genotype are currently available, but prenatal diagnosis is dependent upon obtaining specimens of fetal blood, something which has never been attempted. With the development of the "fetal amnioscope" drawing of fetal blood may become a possibility.

If small amounts of fetal blood could be drawn, red cells from the fetus which normally manufacture Hemoglobin F could be "tricked" into synthesizing adult hemoglobin by inhibiting production of normal fetal hemoglobin. Once fetal hemoglobin is inhibited, the red cells would be compelled to produce Hemoglobin A and/or Hemoglobin S, indicating a propensity towards the disease. Using chromatography, cells could then be separated and an evaluation made.

Gene Therapy for Human Genetic Disease?

Theodore Friedmann and Richard Roblin

At least 1500 distinguishable human diseases are already known to be genetically determined (*1*), and new examples are being reported every year. Many human genetic diseases are rare. For example, the incidence of phenylketonuria is about one per 18,000 live births or about 200 to 300 cases per year in the United States (*2*). Others, such as cystic fibrosis of the pancreas, occur about once in every 2500 live births (*3*). When considered together as a group, however, genetic diseases of humans are becoming an increasingly visible and significant medical problem, at least in the developed countries. While the molecular basis for most of these diseases is not yet understood, a recent review (*4*) listed 92 human disorders for which a genetically determined specific enzyme deficiency has been identified.

Concurrent with the recent progress toward biochemical characterization of human genetic diseases have been the dramatic advances in our understanding of the structure and function of the genetic material, DNA, and our ability to manipulate it in the test tube. Within the last 3 years, both the isolation of a piece of DNA containing a specific group of bacterial genes (*5*) and the complete chemical synthesis of the gene for yeast alanine transfer RNA (*6*) have been reported. These advances have led to proposals (*7*) that exogenous "good" DNA be used to replace the defective DNA in those who suffer from genetic defects. In fact, a first attempt to treat patients suffering from a human genetic disease with foreign DNA has already been made (*8*).

Nevertheless, we believe that examination of the current possibilities for DNA-mediated genetic change in humans in the light of some of the requirements for an ethically acceptable medical treatment raises difficult questions. In order to focus the discussion, in this article we concentrate on the prospects for using isolated DNA segments or mammalian viruses as vectors in gene therapy. For this reason we do not discuss other techniques, such as cell hybridization (*9*), which have been used to introduce new genetic material into mammalian cells. We limit our discussion to the possible therapeutic uses of genetic engineering in humans. The potential eugenic uses, for example, the improvement of human intelligence or other traits, are not discussed because they will be very much more difficult to accomplish (*10*) and raise rather different ethical questions. Whether genetic engineering techniques can be developed for therapeutic purposes in human patients without leading to eugenic uses is an important question, but lies mostly beyond the scope of this article.

Schematic Model of Genetic Disease

Some aspects of a hypothetical human genetic disease in which an enzyme is defective are shown in Fig. 1. The consequences of a gene mutation which renders enzyme E_3 defective could be (i) failure to synthesize required compounds D and F; (ii) accumulation of abnormally high concentrations of compound C and its further metabolites by other biochemical pathways; (iii) failure to regulate properly the activity of enzyme E_1, because of loss of the normal feedback inhibitor, compound F; and (iv) failure of a regulatory step in a linked pathway because of absence of compounds D or F, as in the increased synthesis of ketosteroids in the adrenogenital syndrome (11). In some cases of human genetic disease, accumulation of high concentrations of compound C and its metabolites appears to do the damage. Often a consequence is mental retardation.

The pathway in Fig. 1 is typical of some recessively inherited genetic defects which result in a deficiency of some gene product, usually an enzyme or hormone. In theory, such defects might be corrected by gene therapy, since such techniques might be able to restore the deficient gene product. Other kinds of genetic defects, including those such as the Marfan syndrome which show dominant inheritance and those such as Mongolism that are caused by chromosome abnormalities, could probably not be ameliorated by the kind of gene therapy we emphasize here.

Current Therapy

Human genetic diseases are usually treated by dietary therapy (12), drug therapy, or gene product replacement therapy (11). For example, diets low in lactose or phenylalanine are used as treatments for individuals with galactosemia and phenylketonuria, respectively. Such diets have proved exceedingly effective in galactosemia and have produced a marked reduction in the incidence of mental retardation associated with phenylketonuria. In terms of Fig. 1 this therapy corresponds to restricting the intake of compound A, thus minimizing the accumulation of compound C whose further normal metabolism is blocked.

Drug therapy has been used to block or reduce the accumulation of undesired and possibly harmful metabolites. One example is the inhibition of the enzyme xanthine oxidase with the drug allopurinol to reduce the accumulation of uric acid associated with gout and the Lesch-Nyhan syndrome (13). At present, this method of treatment has been applied to only a few human genetic diseases. Its more general application clearly depends upon the availability of drugs which act selectively on specific enzymes. In another form of drug therapy, drugs which combine specifically with the accumulated compound C are used. An example is the use of D-penicillamine to promote excretion of excess cystine in patients with cystinuria (11).

In theory, some human genetic diseases might be alleviated by supplying directly the deficient enzyme (E_3 in Fig. 1). Recently, attempts to treat Fabry's disease (14), metachromatic leukodystrophy (15), and type 2 glycogenosis (16), by administering the missing enzyme have been reported. Since exogenous enzyme molecules are eventually inactivated or excreted from the body, repeated enzyme injections would be required to manage the diseases in this way. In time, the patients would probably respond by forming antibodies

GENE G_1 G_2 G_3 G_4

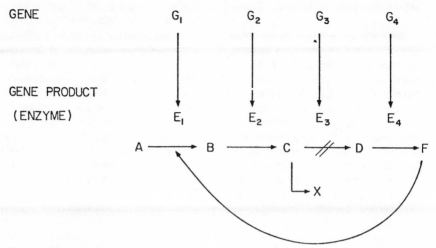

Fig. 1. A hypothetical pathway for the enzymatic conversion of compound A to a final metabolic product F. Compounds B, C, and D are intermediate products. Four different enzymes, E_1, E_2, E_3, and E_4, the products of the corresponding genes G_1, G_2, G_3, and G_4 are required to effect the conversion. The block occurs in the conversion of compound C to D. The concentration of compound F regulates the activity of the first enzyme in the pathway, E_1, in a feedback control loop.

against the administered enzyme. However, insoluble or encapsulated enzyme preparations may in the future provide a means of supplying therapeutic enzymes in a more stable and perhaps less immunogenic form.

There are growing possibilities for the early detection of some genetic diseases by diagnosis in utero. It is now possible to sample the cells of a growing fetus in utero and by examining these cells to diagnose a variety of genetic defects (17). If genetic defects are detected, some states will permit an abortion if the prospective parents so desire. We recognize that diagnosis in utero and abortion raise difficult social and ethical problems of their own and cite them only to indicate that there are additional alternatives to prospective gene therapy for coping with human genetic disease.

However, many genetic diseases do not yet respond to any of these treatments. For example, most genetic disorders of amino acid metabolism (other than phenylketonuria) cannot be well controlled by dietary therapy. Storage diseases associated with lysosomal enzyme deficiencies (18) do not appear to respond to enzyme therapy (14, 15) and will probably be impossible to control by dietary restriction. In addition, even in cases where disease management is effective, it is seldom perfect. Individuals with diabetes mellitus, when treated with insulin, have an increased incidence of vascular and other disorders and a decreased life expectancy compared to the nondiabetic population (19). Children with Lesch-Nyhan syndrome may have their uric acid accumulation controlled by drug (allopurinol) therapy, but their brain dysfunction has, to date, not been reversible.

These limitations of current therapy are stimulating attempts to develop techniques for treating human genetic diseases at the genetic level, the site of the primary defect. Genetic modification of specific characteristics of human cells by means of exogenous DNA seems possible for several reasons. DNA-mediated genetic modification of several different kinds of bacteria has been known for many years (20), and recent experiments suggest that the genetic properties of mutant *Drosophila* strains can be modified by treating their eggs with DNA extracted from other *Drosophila* strains (21). It has also been found that treatment of human cells in vitro with DNA extracted from the oncogenic virus SV40 results in permanent heritable alteration of several cellular properties (22).

Genetic Modification Mediated by DNA

Permanent, heritable, genetic modification of a human cell by means of DNA requires (i) uptake of the exogenous DNA from the extracellular environment; (ii) survival of at least a portion of the DNA during its intracellular passage to the nucleus; (iii) stabilization of the exogenous DNA in the recipient cell; and (iv) expression of the new genes via transcription into an RNA message (mRNA) and translation of this message into the appropriate protein. Some of these processes are illustrated schematically in Fig. 2.

Mammalian cells take up proteins, nucleic acids, and viruses from their environment by a process known as endocytosis (23). After binding to the cell membrane, the macromolecules are drawn into the cell by an infolding of the external cell membrane leading to vesicle formation (see Fig. 2). Macromolecules contained in vesicles derived from invaginations of the external cell membrane can be degraded if these vesicles fuse with lysosomes. Lysosomes are cell organelles which contain a variety of hydrolytic enzymes. These enzymes can rapidly degrade ingested macromolecules, including DNA (24). Thus, mammalian cells possess mechanisms for protecting themselves from the potentially perturbing influences of foreign DNA.

Despite this cellular defense mechanism, exogenous foreign DNA can, under certain circumstances, become integrated in the DNA of the recipient cell. The evidence of this has come from studies of oncogenic virus transformation of mammalian, including human, cells (25). In the case of oncogenic transformation with SV40 virus, the viral DNA is apparently physically integrated into the chromosomal DNA of the recipient cell (26). It seems probable that heritable alterations of cell morphology and biochemistry are the result of the expression of one or more viral genes. Presumably, viral DNA integration takes place by base pairing of homologous regions of host cell and viral DNA followed by genetic recombination. However, the integration of oncogenic viral DNA may represent a special case since at least one viral gene product may be required for integration (27). Other, nonviral DNA molecules, unable to supply this integration function, might integrate at a much lower frequency, if at all.

In addition to integration by genetic recombination, exogenous DNA might be stabilized in the recipient cell as an independently replicating genetic unit in the cell nucleus. Although such units are known to exist in bacteria, they have not been observed in mammalian cells. However, the cytoplasmic mitochondria of mammalian cells do contain nonchromosomal, independently replicating units of DNA. The mitochondrial

DNA replication system thus offers another possible site for stabilization of exogenous DNA.

For a human genetic defect to be repaired by administering exogenous DNA, the stabilized newly introduced DNA must be correctly expressed. That is, the new gene must be correctly transcribed into mRNA and this mRNA must be correctly translated into protein. Since little is known about the regulation of mRNA synthesis and translation during natural gene expression in mammalian cells, a corresponding high degree of uncertainty exists concerning the ability of newly introduced DNA to be expressed correctly.

A variety of attempts have been made to demonstrate DNA-mediated modification of genetically mutant mammalian cells, both in vivo and in vitro. Apparently successful results in vitro have been reported for diploid human cells lacking the purine "salvage" enzyme hypoxanthine-guanine phosphoribosyltransferase (HGPRT); in human reticulocytes synthesizing an abnormal hemoglobin; in several malignant cells of mouse origin carrying markers for drug resistance; and in mouse cells with defective melanin synthesis, among others (28). In addition, transient expression of HGPRT enzyme function has been detected in human cells deficient in HGPRT after exposing them to DNA from cells with normal amounts of HGPRT (29). This suggests that exogenous DNA may be taken up and expressed, without necessarily being stabilized. However, none of the successful experiments described to date have been reproducible.

There may be several reasons for failure to demonstrate consistently the genetic modification of mammalian cells by DNA. Many previous experiments suffered from the unavailability of good genetic markers and sensitive selective systems for detecting modified cells. An important difficulty in using bulk DNA isolated from human (or other mammalian) cells is that the fraction of this DNA which is specific for any given gene is estimated to be extremely small, of the order of 10^{-7} (30). As we mentioned earlier, nonviral exogenous DNA may not be able to integrate into the chromosomal DNA of the recipient cell, thus preventing permanent genetic modification.

In spite of the lack of reproducible success in past experiments, several recent technical developments have suggested new ways in which the problems of low DNA specificity, failure of integration, and intracellular DNA degradation might be overcome.

The prospects for directing genetic modification of mammalian cells would almost certainly be enhanced by using DNA preparations containing only the gene for which the genetically defective cells are mutant. As already pointed out, both the isolation of a specific group of bacterial genes and the complete chemical synthesis of a single gene were reported recently (5, 6).

The RNA-dependent DNA polymerase recently found in RNA tumor viruses (31) could also be used for gene synthesis in vitro. Since this enzyme is able to make DNA copies from an RNA template, it offers a method for synthesizing the DNA for any specific RNA which might be isolated in pure form. Thus, it seems probable that our developing ability to isolate specific genes, or synthesize them, will eventually eliminate the problem of low specificity of the exogenous DNA.

Some workers are developing techniques which could be used to overcome the problem of stabilizing the incoming exogenous DNA in the recipient cell (32). They plan to make use of the ability of the DNA from SV40 virus to integrate into the chromosomal DNA of the cell. Specific genes will be attached to the viral DNA by means of several biochemical steps which are already known and fairly well characterized. These operations would create a hybrid DNA molecule which would carry the information for integration from the original viral DNA and perform the specific gene functions of the attached DNA. In this approach, DNA integration would be combined with biochemical manipulation of the DNA gene substance in vitro, and any gene-specific DNA segments obtained by synthesis or isolation could be utilized. It is clear that before such hybrid DNA molecules could be used in a human therapeutic situation, the oncogenic potential of the viral DNA would have to be eliminated.

In another experimental approach, virus-like particles which contain pieces of cellular DNA (pseudovirions) instead of viral DNA are being used as the vector for DNA-mediated genetic modification (*33*). This might help to protect the incoming DNA from intracellular degradation. However, pseudovirion DNA is probably a random collection of cellular DNA fragments (*34*) and hence nonspecific for any given gene; it might also be unable to stabilize itself by integration into the host cell DNA. This may explain why attempts to modify thymidine kinase–deficient mouse cells in vitro by means of polyoma virus pseudovirions have been unsuccessful (*35*).

It has recently proved possible to reconstruct infectious particles of several plant and bacterial viruses from the nu-

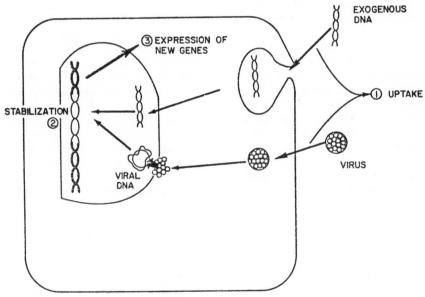

Fig. 2. Steps in the genetic modification of a mammalian cell. The added exogenous genetic information may be integrated into the chromosome of the recipient cell and become expressed as a new gene product.

cleic acid and capsid protein components (*36*). This suggests the possibility of creating artificial pseudoviruses as vectors for DNA-mediated genetic modification. These pseudovirions would contain specific DNA segments (either isolated or synthesized) surrounded by virus capsid protein. The probability of introducing a specific piece of genetic material might be greatly increased when compared with natural pseudovirions carrying randomly excised pieces of DNA. This in no way solves the difficulties of integration and expression of the genetic material. Since the specificity of virus-cell interactions is determined at least in part by the virus capsid protein, encapsidation of specific DNA molecules might confer some cell or tissue selectivity upon the DNA molecules used for gene modification.

DNA-Mediated Gene Therapy

In attempting to envision how DNA might be used as a mediator for the modification of genes in a human being suffering from a genetic defect, we foresee several kinds of new problems. First, the existence of differentiation and cell specialization in the human body will pose several questions. Many human genes are active or expressed only in a small fraction of the cells of the body. For example, the activity of the enzyme phenylalanine hydroxylase (deficient in individuals with phenylketonuria) is demonstrable only in the liver. For prospective gene therapy there might be several consequences. (i) The introduction of, for example, the gene for phenylalanine hydroxylase into cells which do not normally express this enzyme would yield no therapeutic benefits if the expression of the newly introduced genes were also blocked. Methods would have to be developed to deliver the exogenous DNA

to the appropriate "target tissue," and to confine its action solely to that tissue. (ii) Some gene products (hormones, for example) are made and secreted by one specialized group of cells and act on target cells elsewhere in the body. Synthesis and secretion of hormones such as insulin are regulated by mechanisms which are still imperfectly understood. Thus, the introduction of new genes for insulin into cells not appropriately differentiated to provide the correct synthetic and secretory responses would be of little use as a treatment for diabetes. (iii) In several genetic disorders, genetic modification of the brain cells themselves may be required to reduce the accumulation of metabolites in the brain, because the blood-brain barrier might prevent enzymes made in other parts of the body from entering the brain (*15*). We wonder whether direct genetic modification of brain cells could be made safe enough for use in human patients.

Second, regulation of the quantitative aspects of enzyme production may present a problem. By mechanisms as yet unknown, concentrations of cellular enzymes are regulated so that neither too much nor too little enzyme is produced by normal cells. How will we ensure that the correct amount of enzyme will be made from the newly introduced genes? Will the integration event, linking exogenous DNA to the DNA of the recipient cell, itself disturb other cellular regulatory circuits?

Third, the patient's immunological system must not recognize as foreign the enzyme produced under the direction of the newly introduced genes. If this occurred, the patient would form antibodies against the enzyme protein, perhaps nullifying the intended effects of the genetic intervention. This suggests that the new gene introduced during gene therapy would have to code for an enzyme with the same amino acid sequence as the human enzyme.

In addition, administration of foreign genetic material to patients carries a risk of altering the germ cells as well as the desired target cells. One might think that this problem could be circumvented by first removing some of the patients' cells, carrying out DNA-mediated genetic modification in vitro, and then reimplanting the altered cells back into the patient. However, this approach is likely to be limited by the tendency of cells to dedifferentiate and become malignant when grown in vitro.

For an acceptable genetic treatment of a human genetic defect, we would require that the gene therapy replace the functions of the defective gene segment without causing deleterious side effects either in the treated individual or in his future offspring. Years of work with tissue cultures and in experimental animals with genetic defects will be required to evaluate the potential side effects of gene therapy techniques. In our view, solutions to all these problems are needed before any attempt to use gene therapy in human patients could be considered ethically acceptable.

We are aware, however, that physicians have not always waited for a complete evaluation of new and potentially dangerous therapeutic procedures before using them on human beings. Consider how little was known of the basic aspects of virology during Jenner's development of vaccination against smallpox. In this regard, potential gene therapy techniques resemble other medical innovations. There is currently, and there may continue to be, a tendency to use incompletely understood genetic manipulative techniques, borrowed from molecular biology, in clinical settings. We believe that the first attempt at gene therapy in human patients (8) illustrates this con-

tention.

The case in question (8) concerns two children suffering from hyperargininemia, a hereditary deficiency of the enzyme arginase. The arginase deficiency leads to high concentrations of arginine in the children's blood and cerebrospinal fluid, and has associated with it severe mental retardation (37). An attempt has been made to correct this defect at the genetic level by injecting Shope papilloma virus into the children (8). The scientific rationale for this treatment is based upon the report that the synthesis of arginase is stimulated in rabbit skin infected with Shope papilloma, and that this new arginase activity had some properties which are different from those of the normal enzyme of rabbit liver (38). In 1958, when these experiments were first reported, it was postulated that the viral DNA carried the gene for a viral arginase different from the cellular enzyme. In addition, the serums of laboratory workers who had worked with and thus been exposed to Shope papilloma virus were tested, and 35 percent of them exhibited lower concentrations of arginine than control hospital patients who had not knowingly been exposed to the virus (39). Thus, there were some grounds for believing that inadvertent infection with Shope papilloma in humans could lower the concentration of serum arginine without apparent harmful effects.

More recently, the interpretation that Shope papilloma virus codes for an arginase has been seriously questioned (40). It now appears more probable that the virus infection stimulates the production of a cellular arginase. Whether the induced arginase is coded for by viral or by cellular genes is important to the rationale of this attempt at gene therapy. If virus infection induces the synthesis of cellular arginase, and if the children have hereditarily lost the ability to produce arginase, then infecting the children with Shope papilloma virus may not have any possibility of correcting their condition (41).

The use of intact viruses as vectors in gene therapy raises further questions. When applied to the skin of rabbits Shope papilloma virus induces skin papillomas, a variable proportion of which develop into cancerous skin lesions. Although Shope papilloma has not had any known harmful effects on humans, tests to establish the safety of large doses have not been performed. It should also be shown that a vector for clinical gene therapy is free from other contaminating viruses latent in the cells used to produce the injected virus.

The clinical results of this therapeutic attempt are not yet known. But we are concerned that this first attempt at gene therapy, which we believe to have been premature, will serve as an impetus for other attempts in the near future. For this reason, we offer the following considerations as a starting point for what we hope will become a widespread discussion of appropriate criteria for the use of genetic manipulative techniques in humans.

Some Preliminary Criteria

We propose the following ethico-scientific criteria which any prospective techniques for gene therapy in human patients should satisfy:

1) There should be adequate biochemical characterization of the prospective patient's genetic disorder. It should be determined whether the patient (i) is producing a mutated, inactive form of the normal protein; (ii) is producing none of the normal protein; or (iii) is producing the normal protein in normal amounts, but the protein is rendered inactive in some way. For example, alterations in membrane structure leading to loss of the cellular receptors for insulin could produce a diabetes-like condition, even though the patient were producing normal amounts of insulin. We anticipate that defects of this type may be found affecting the activity of enzymes which are normally constituents of cell membranes. Our point is that only in the first type of genetic defect (i) would currently envisioned gene therapy techniques be likely to improve the patient's condition.

2) There should be prior experience with untreated cases of what appears to be the same genetic defect so that the natural history of the disease and the efficacy of alternative therapies can be assessed. Thus, the first reported cases of a new human genetic disease would seldom be candidates for attempts at gene therapy. The reason for this criterion comes from our accumulating experience with some of the better studied genetic defects such as phenylketonuria and galactosemia. We now observe heterogeneity in these conditions; that is, what appears to be the same genetic disease turns out to have different genetic bases in different individuals. Widespread screening for phenylketonuria in newborns has detected individuals who, like phenylketonurics, have high concentrations of phenylalanine in the serum just after birth, but have concentrations in the normal range several months later (2). It is now also clear that some individuals with high concentrations of phenylalanine in the serum have normal intelligence quotients (2). We anticipate that other genetic diseases will exhibit the same kind of heterogeneity. Concern for the welfare of each individual patient dictates that we not rush in with gene therapy until we are very sure about

the precise nature and consequences of his genetic defect.

3) There must be an adequate characterization of the quality of the exogenous DNA vector. This will require the development of new, more accurate methods of analyzing the base sequence of the DNA, if synthetic DNA molecules are to be used, or the development of new methods of isolation and purification, if naturally occurring DNA molecules are to be used. We visualize the Food and Drug Administration, or some similar organization, establishing and enforcing quality standards for DNA preparations used in gene therapy.

4) There should be extensive studies in experimental animals to evaluate the therapeutic benefits and adverse side effects of the prospective techniques. These tests should include long-term studies on the possible induction of cancer and genetic disturbances in the offspring of the treated animals. This will require the development of animal models for human genetic diseases. Previous work, which led to the isolation of a mouse strain deficient in the enzyme catalase (42) suggests that such animal models could be developed and might yield answers to some of the questions we have raised.

5) For some genetic diseases, the patient's skin fibroblasts grown in vitro reflect the disorder. Thus, in some cases it would be possible to determine whether the prospective gene therapy technique could restore enzyme function in the cells of the prospective patient. This could be done first in vitro, without any of the risks of treating the whole patient. Some side effects, such as chromosome damage and morphological changes suggesting malignancy, could also be assessed at this time. Only when a potential gene therapy technique had satisfied all these safety and efficacy criteria would it be considered for use in human patients.

These criteria omit some other considerations which we believe are important. Although the ethical problems posed by gene therapy are similar in principle to those posed by other experimental medical treatments, we feel that the irreversible and heritable nature of gene therapy means that the tolerable margin of risk and uncertainty in such therapy is reduced. Physicians usually arrive at a judgment regarding the ethical acceptability of an experimental therapy by balancing the risks and consequences of different available treatments against their potential benefits to the patient. In general, the degree of risk tolerated in medical treatment is directly related to the seriousness of the condition.

High-risk treatments are sometimes considered more justified in life-threatening situations. For different human genetic diseases, the severity of the problem in the untreated condition and the response to currently available therapy varies greatly. Thus, phenylketonuria leads to mental retardation, but not death, in most untreated affected individuals, but the mental retardation can be avoided for the most part by prompt neonatal dietary therapy. In contrast, in the infantile form of Gaucher's disease, a deficiency in the enzyme glucocerebrosidase (important in the metabolism of brain glycolipids) leads to severe and progressive neurologic damage and death within 1 or 2 years (38). There is as yet no effective therapy. Thus, the specific characteristics of each genetic disease will be an important factor in evaluating whether or not to attempt gene therapy. We believe that the prospective use of gene therapy will need to be evaluated on a case by case basis.

Another ethical ideal which guides experimental medical treatments is informed consent. By informed consent we mean that the patient, after having the nature of the proposed treatment and its known and suspected risks explained to him by the physician, freely gives the physician his consent to proceed with the treatment. Since many of the cases where gene therapy might be indicated will involve children or newborns as patients, there will be especially troubling problems surrounding informed consent. Parents of newborn children with genetic defects may be asked to give "consent by proxy" for gene therapy. Clearly, until we know much more about the side effects of gene therapy, it will not be possible to provide them with adequate information about risks to the treated individual and his offspring.

Control of Gene Therapy

How can gene therapy in humans be controlled to avoid its misuse? By misuse we mean the premature application of techniques which are inadequately understood and the application of gene therapy for anything other than for the primary benefit of the patient with the genetic disease. In our view, it will be possible to control the procedures used for gene therapy at several levels. For example, between the patient and physician, we can usually rely upon the selection of a therapeutic technique having optimal chances of success. In general, we believe that the doctor will not recommend and the patient will not accept an uncertain, risk-laden gene therapy if a reasonably effective alternative therapy is available. However, the physician, in this as in other cases of experimental therapeutic techniques, has a near monopoly on the relevant facts about risks and benefits of various treatments. Since the physician concerned may also be active in

trying to develop the gene therapy technique, how can the patient be protected from a physician who might be over-eager to try out his new procedure?

It seems to us that significant opportunities for control also exist at the level of the hospital committees responsible for examining experimental techniques. Already at accredited hospitals, all proposals for research in which human subjects will be used must pass through a review committee. Further control exists through scrutiny of the proposed techniques by the physician's immediate peers.

Procedures to be used for gene therapy might also be controlled by the committees and organizations approving and funding research grants. Moderately large amounts of money will be required for the development of gene therapy techniques, hence there should be competition for public funds with other urgent medical needs. Thus, the first use of gene therapy in human patients would, of necessity, have secured the implied or direct approval of several larger public bodies beyond the principal physician-investigator. In our judgment, these levels of control will probably prove adequate to prevent misuse of projected gene therapy if, as we suspect, gene therapy is attempted in only a small number of instances. Any potential large-scale use of gene therapy (for example, the prospect of treating the approximately 4 million diabetics in the United States with DNA containing the gene for insulin) might appreciably affect the overall quality of the gene pool and would require other forms of control.

Conclusions

In our view, gene therapy may ameliorate some human genetic diseases in the future. For this reason, we believe that research directed at the development of techniques for gene therapy should continue. For the foreseeable future, however, we oppose any further attempts at gene therapy in human patients because (i) our understanding of such basic processes as gene regulation and genetic recombination in human cells is inadequate; (ii) our understanding of the details of the relation between the molecular defect and the disease state is rudimentary for essentially all genetic diseases; and (iii) we have no information on the short-range and long-term side effects of gene therapy. We therefore propose that a sustained effort be made to formulate a complete set of ethico-scientific criteria to guide the development and clinical application of gene therapy techniques. Such an endeavor could go a long way toward ensuring that gene therapy is used in humans only in those instances where it will prove beneficial, and toward preventing its misuse through premature application.

Two recent papers have provided new demonstrations of directed genetic modification of mammalian cells. Munyon *et al.* (*44*) restored the ability to synthesize the enzyme thymidine kinase to thymidine kinase–deficient mouse cells by infection with ultraviolet-irradiated herpes simplex virus. In their experiments the DNA from herpes simplex virus, which contains a gene coding for thymidine kinase, may have formed a hereditable association with the mouse cells. Merril *et al.* (*45*) reported that treatment of fibroblasts from patients with galactosemia with exogenous DNA caused increased activity of a missing enzyme, α-D-galactose-1-phosphate uridyltransferase. They also provided some evidence that the change persisted after subculturing the treated cells. If

this latter report can be confirmed, the feasibility of directed genetic modification of human cells would be clearly demonstrated, considerably enhancing the technical prospects for gene therapy.

References and Notes

1. V. A. McKusick, *Mendelian Inheritance in Man* (Johns Hopkins Press, Baltimore, ed. 3, 1971).
2. D. Y. Y. Hsia, *Progr. Med. Genet.* **7**, 29 (1970).
3. E. R. Kramm, M. M. Crane, M. G. Sirken, M. D. Brown, *Amer. J. Public Health* **52**, 2041 (1962).
4. V. A. McKusick, *Annu. Rev. Genet.* **4**, 1 (1970).
5. J. Shapiro, L. MacHattie, L. Eron, G. Ihler, K. Ippen, J. Beckwith, *Nature* **224**, 768 (1969).
6. K. L. Agarwal, H. Buchi, M. H. Caruthers, N. Gupta, H. G. Khorana, K. Kleppe, A. Kumar, E. Ohtsuka, U. L. Rajbhandary, J. H. Van De Sande, V. Sgaramella, H. Weber, T. Yamada, *ibid.* **227**, 27 (1970).
7. S. Rogers, *New Sci.* (29 Jan. 1970), p. 194; H. V. Aposhian, *Perspect. Biol. Med.* **14**, 98 (1970).
8. *New York Times*, 20 Sept. 1970.
9. A. G. Schwartz, P. R. Cook, H. Harris, *Nature* **230**, 5 (1971).
10. B. D. Davis, *Science* **170**, 1279 (1970).
11. J. B. Stanbury, J. B. Wyngaarden D. S. Fredrickson, Eds., *The Metabolic Basis of Inherited Disease* (McGraw-Hill, New York, ed. 2, 1966).
12. N. A. Holtzman, *Annu. Rev. Med.* **21**, 335 (1970).
13. W. N. Kelley, F. M. Rosenbloom, J. Miller, J. E. Seegmiller, *N. Engl. J. Med.* **278**, 287 (1968).
14. C. A. Mapes, R. L. Anderson, C. C. Sweeley, R. J. Desnick, W. Krivit, *Science* **169**, 987 (1970).
15. H. L. Greene, H. Hug, W. K. Schubert, *Arch. Neurol.* **20**, 147 (1969).
16. G. Hug and W. K. Schubert, *J. Cell Biol.* **35**, C1 (1967).
17. A. Milunsky, J. W. Littlefield, J. N. Kanfer, E. H. Kolodney, V. E. Shih, I. Atkins, *N. Engl. J. Med.* **283**, 1370, 1441, 1498 (1970).
18. R. O. Brady, *Annu. Rev. Med.* **21**, 317 (1970).
19. S. Pell and C. A. D'Alonzo, *J. Amer. Med. Ass.* **214**, 1833 (1970).
20. R. D. Hotchkiss and M. Gabor, *Annu. Rev. Genet.* **4**, 193 (1970).
21. A. S. Fox, S. B. Yoon, W. M. Gelbart, *Proc. Nat. Acad. Sci. U.S.* **68**, 342 (1971).
22. S. A. Aaronson and G. J. Todaro, *Science* **166**, 390 (1969).
23. H. J. P. Ryser, *ibid.* **159**, 390 (1968).
24. J. T. Dingle and H. B. Fell, *Frontiers Biol.* **14A**, 220 (1969).
25. R. Dulbecco, *Science* **166**, 962 (1969); M. Green, *Annu. Rev. Biochem.* **39**, 701 (1970).
26. J. Sambrook, H. Westphal, P. R. Srinivasan,

R. Dulbecco, *Proc. Nat. Acad. Sci. U.S.* **60**, 1288 (1968).

27. M. Fried, *Virology* **40**, 605 (1970).

28. E. H. Szybalska and W. Szybalski, *Proc. Nat. Acad. Sci. U.S.* **48**, 2026 (1962); M. Fox, B. W. Fox, S. R. Ayad, *Nature* **222**, 1086 (1969); R. A. Roosa and E. Bailey, *J. Cell. Physiol.* **75**, 137 (1970); L. M. Kraus, *Nature* **192**, 1055 (1961); D. Roth, M. Manjon, M. London, *Exp. Cell Res.* **53**, 101 (1968); J. L. Glick and C. Sahler, *Cancer Res.* **27**, 2342 (1967); E. Ottolenghi-Nightingale, *Proc. Nat. Acad. Sci. U.S.* **64**, 184 (1969).

29. T. Friedmann, J. H. Subak-Sharpe, W. Fujimoto, J. E. Seegmiller, paper presented at Society of Human Genetics Meeting, San Francisco, Oct. 1969.

30. Calculated with the assumption that (i) the DNA contents of a diploid human cell is 4×10^{12} daltons; (ii) representative gene codes for a protein containing 200 amino acids equivalent to 4×10^5 daltons of DNA; and (iii) there are two copies of each gene per cell. This would represent a minimum estimate if the redundant DNA segments in human cells include genes which specify enzymes.

31. H. M. Temin and S. Mizutani, *Nature* **226**, 1211 (1970); D. Baltimore, *ibid.*, p. 1209.

32. P. Berg, and D. M. Jackson, personal communication.

33. J. V. Osterman, A. Waddell, H. V. Aposhian, *Proc. Nat. Acad. Sci. U.S.* **67**, 37 (1970).

34. L. Grady, D. Axelrod, D. Trilling, *ibid.*, p. 1886.

35. B. Hirt, seminar at Brandeis University, Oct. 1970.

36. H. Fraenkel-Conrat, *Annu. Rev. Microbiol.* **24**, 463 (1970).

37. H. G. Terheggen, A. Schwenk, A. Lowenthal, M. Van Sande, J. P. Colombo, *Lancet* **II-1969**, 748 (1969); Z. *Kinderheilk* **107**, 298 and 313 (1970).

38. S. Rogers, *Nature* **183**, 1815 (1959).

39. —— and M. Moore, *J. Exp. Med.* **117**, 521 (1963).

40. P. S. Satoh, T. O. Yoshida, Y. Ito, *Virology* **33**, 354 (1967); G. Orth, F. Vielle, J. B. Changeux, *ibid.* **31**, 729 (1967).

41. This objection would be straightforward if only one type of arginase were present in human cells. Even though there seem to be two arginase isozymes in both human liver and erythrocytes, the objection is still cogent since one isozyme contributes 90 to 95 percent of the total arginase activity, and the two isozymes cross-react immunologically. See J. Cabello, V. Prajoux, M. Plaza, *Biochim. Biophys. Acta* **105**, 583 (1965).

42. R. N. Feinstein, M. E. Seaholm, J. B. Howard, W. L. Russel, *Proc. Nat. Acad. Sci. U.S.* **52**, 661 (1964).

43. H. Harris, *Frontiers Biol.* **19**, 167 (1970).

44. W. Munyon, E. Kraiselburd, D. Davis, J. Mann, *J. Virol.* **7**, 813 (1971).

45. C. R. Merril, M. R. Geier, J. C. Petricciani, *Nature* **233**, 398 (1971).

110

Molecular Biology:
Gene Insertion into Mammalian Cells

By DANIEL RABOVSKY

The problem of inserting specific genes into human cells has intrigued molecular geneticists, and the prospect of the successful solution of this problem has concerned everyone. Both the excitement and the concern have grown now that the armchair speculations— and exploratory results—of a few years ago have matured into hard experimental work. The current results of that work indicate that animal viruses, bacterial viruses, and cell fusion techniques are all capable of introducing new functional genes into mammalian cells, although many of the fundamental genetic and regulatory processes in mammalian cells remain unknown.

Much has been learned about the genetic code and the mechanisms of the replication of DNA, the transcription of DNA into RNA, and the translation of RNA into protein, especially in bacterial cells. A clever and sufficiently industrious molecular geneticist can often produce a specific mutation in any of a large number of genes in the bacterium *Escherichia coli*, can delete genes or add new ones from outside the cell, and can then regulate the expression of genetic traits inside the cell. But the extension of these techniques from bacteria and bacterial viruses (bacteriophages) to nucleated (eukaryotic) cells, especially human cells, awaited new tools and more knowledge.

Several biologists have studied the interaction of foreign DNA with nucleated cells. Among these, Pradman Qasba and Vasken Aposhian at the University of Maryland School of Medicine in Baltimore, have recently shown that one type of animal virus can be used to transport DNA from mouse cells into the nuclei of human cells. At the Roswell Park Memorial Institute in Buffalo, W. Munyon and his coworkers have shown that another type of animal virus may have inserted a specific gene into mouse cells without harming the cells. These workers found that the enzyme specified by this gene was made by the cell and that the new gene seemed to be replicated as the cells divided.

Munyon and his group infected mutant L cells (a line of mouse tissue culture cells) that lacked the enzyme thymidine kinase with the animal virus

SCIENCE, 1971, Vol. 174, pp. 933-934. Copyright 1971 by the American Association for the Advancement of Science.

herpes simplex. The virus had been irradiated with ultraviolet light to decrease its ability to kill cells (*1*). Herpes simplex virus normally induces a thymidine kinase activity during infection before it kills the cells, but in this experiment about 0.1 percent of the infected L cells were transformed by the irradiated virus into stable cells that had thymidine kinase activity and were maintained in culture for 8 months. No measurable proportion ($< 10^{-8}$) of control L cells gained the ability to express thymidine kinase when uninfected cells or cells infected with a herpes simplex mutant that does not induce thymidine kinase activity were examined.

These results are consistent with the idea that the herpes simplex virus introduced a gene for thymidine kinase into the L cells and that this gene was then maintained and replicated by the cells. However, Munyon notes the possibility that a herpes gene product may have simply induced the stable expression of a gene that was already present in the L cells.

Aposhian has proposed that pseudovirions—which consist of normal virus' protein coats that have enclosed foreign pieces of DNA—might be able to deliver foreign genes into another cell, and that these new genes could function and be replicated in the cell. Qasba and Aposhian have now established that pseudovirions of an animal virus, polyoma, containing labeled DNA from mouse embryo cells can deliver this DNA inside the nuclei of human embryo cells in the form of uncoated pseudovirion DNA as soon as 24 hours after infection of the human cells by the polyoma pseudovirions (*2*). At present no actual expression or replication of any newly introduced genes has been shown in this system, but many workers think it likely that uncoated mammalian DNA in a mammalian cell nucleus is capable of integrating itself into the host chromosome and functioning in some way.

Carl Merril and his co-workers (*3*) at the National Institutes of Health in Bethesda have taken a different approach. They have shown that a bacteriophage is capable of introducing a selected functional gene into human cells. Merril had worked with the bacterial virus lambda phage, one of the transducing phages of *E. coli*. Transducing phages have been among the most useful genetic tools available to molecular biologists. Sometimes bacterial genes are included in the DNA of the new phage produced during infection of a bacterium by a transducing phage. If these phages infect another bacterium they can transduce the bacterial genes that they carry into their new host. In this way transducing phages are often used to selectively introduce new functional genes into bacteria.

The genetic structure of lambda has been very well mapped, and a number of remarkable lambda transducing phages (that is, lambda that carry bacterial genes) are available. Merril's group decided to attempt the transduction of cultured human fibroblast cells from the skin of a patient with galactosemia, the disease that results from an inborn error of metabolism in which the enzyme α-D-galactose-1-phosphate uridyl (GPU) transferase is lacking. The cultured fibroblasts were treated with lambda phage (λ pgal) that carried the *E. coli* galactose operon—a set of genes that codes for the enzymes which convert galactose to glucose and controls their synthesis.

Merril and his co-workers hoped that the GPU transferase gene, which is part

112

of the galactose operon, could be supplied to the fibroblasts by λ pgal and that the fibroblasts would use this transduced gene to make the GPU transferase enzyme.

These expectations were fulfilled. Infection of the fibroblasts by λ pgal resulted in both the production of lambda-specific RNA and in the appearance of GPU transferase activity in the fibroblasts. The lambda-specific RNA and the new GPU transferase activity were found in significant amounts and persisted at the same amounts per cell for more than 40 days (during which more than eight doublings of the cells took place). In addition, uncoated λ pgal DNA was, in Merril's experiment, at least as effective as the whole virus particle. Control infections by normal lambda and by λ pgal with a mutation that inactivates its transferase resulted in the production of lambda-specific RNA but not in transferase activity.

There is still no evidence to indicate what part of the cell houses the lambda DNA. But the experiments performed by Merril's group have shown that lambda DNA can enter human cells, and once there at least some of its genes can be replicated, transcribed, and translated.

Cell Fusion Techniques

Viruses are not the only means of introducing new genes into mammalian cells, however. Henry Harris' group at Oxford in England has now shown that cell fusion can also be used to insert a functional gene from chicken cells into mouse tissue culture cells (4). If two cells are fused or if another nucleus is introduced into a cell, the dividing time of the resulting multinucleate cell is primarily determined by the nucleus that is closest to division. The other nuclei may be forced to divide before they are ready, and their chromosomes usually undergo a "premature condensation" which results in their fragmentation or "pulverization."

The Harris group fused chick red blood cells, whose nuclei carry the gene for chick inosinic acid pyrophosphorylase, with mouse fibroblast A_9 cells, which are a mutant line of mouse L cells deficient in this enzyme. During cell division, the nuclear membranes in these hybrid cells disappear, and the chick chromosomes undergo pulverization. Although only mouse nuclei appeared in the daughter cells after mitosis, some (2×10^{-5}) of the daughter cells had gained the ability to synthesize chick inosinic acid pyrophosphorylase. However, after more than 100 generations in culture, 20 percent of these transformed hybrid cells lost the ability to make this enzyme, as compared to 1 out of 10^6 in a normal L cell population. Hence the chick gene for inosinic acid pyrophosphorylase was replicated and remained functional in the mouse cells, although it was not as genetically stable as the normal mouse gene for this enzyme. This experiment did not provide direct evidence for the location of the chick gene inside the mouse cells, but indirect evidence leads the Harris group to name the mouse nucleus as its probable residence.

A number of laboratories are already extending these gene transfer techniques. Aposhian's group is now studying the genetic properties of polyoma pseudovirions in mice. They are also attempting to produce polyoma pseudovirions containing the gene for human thymidine kinase. Merril has infected whole animals with transducing lambda bacteriophage in order to determine whether any new traits gained by the animals from the genes that are transported by the phage can be inherited from one generation to the next.

The ability to transfer small segments of DNA from one cell to another gives molecular geneticists the opportunity to map the locations of genes within mammalian chromosomes. In turn, understanding the method by which genes are organized into functional units within the chromosome may reveal how the regulation of gene expression takes place. This regulation may play an important role in the differentiation of animal cells.

Current developments certainly do not yet add up to "genetic engineering"; but there now exists very strong evidence, with a number of different techniques, that experimenters can transfer genes between, and insert them into, mammalian cells—a development that opens to experiment many questions about gene function in animals and in humans. Now that a few of the basic tools of molecular genetics have been extended to mammalian cells, more than a few biologists share Aposhian's concern that science "will give us gene therapy before society is prepared for it."

References

1. W. Munyon, E. Kraiselbrud, D. Davis, J. Mann, *J. Virol.* 7, 813 (1971).
2. P. K. Qasba and H. V. Aposhian, *Proc. Nat. Acad. Sci. U.S.* 68, 2345 (1971).
3. C. R. Merril, M. R. Geier, J. C. Petricciani, *Nature* 233, 398 (1971)
4. A. G. Schwartz, P. R. Cook, H. Harris, *Nature New Biol.* 230, 5 (1971).

THE FUTURE OF ASEXUAL REPRODUCTION

James D. Watson

Several years ago a most remarkable frog grew up in Oxford. Its origin did not lie in the union of a haploid sperm cell with a haploid egg, the fertilization process which ordinarily gives each higher animal a mixture of paternal and maternal genes. Instead this frog arose from an enucleated egg, into which had been inserted a diploid nucleus from the intestinal cell of an adult frog. Microsurgical removal of the maternal nucleus from this egg had denuded it of any genetic material.

But by subsequently gaining a diploid nucleus (as opposed to the haploid form found in a sperm) the egg acquired the chromosome number normally present in a fertilized egg. As such it could be activated to divide, thereby setting into motion the successive embryological stages which culminate in an adult frog.

The genetic origin of this frog was thus very different from that of all previous frogs, one half of whose chromosomes came from the male parent through the sperm, the other half from the female parent which produced the egg. Normal fertilization processes, by combining genetic material from two different parents, always generate progeny uniquely different from either parent.

Clonal reproduction of frogs

In contrast, the Oxford frog derived *all* its genetic material from the individual whose intestinal cell was used as the nuclear donor. The genetic complement of all its diploid somatic cells (as opposed to its haploid sex cells) was thus identical to that in the donor frog. So, in effect, it was an identical twin of the donor frog born some months before. Furthermore, since every adult frog contains millions of cells capable of being used as nuclear sources, the original donor could have served as the genetic parent of thousands of progeny identical to itself.

This type of reproduction is generally referred to as a *clonal* reproduction. (A *clone* is the aggregate of the asexually produced progeny of a single cell; for example, all the descendants of a single bacteria present as a colony upon a Petri dish.) The genetic identicalness of all members of a clone arises from the fact that the normal process of cell division (called mitosis) produces two daughter cells with identical chromosomal complements. The nuclei of the cells found in the frog's intestine are thus identical to those which could be found, say, in its liver or brain.

Cell differentiation

In contrast, the cell division process called meiosis, which generates the sex cells, reduces the chromosome number by half. Only one of each pair of homologous

INTELLECTUAL DIGEST, 1971, Vol. 2, No. 2, pp. 69-74.

chromosomes enters a sperm or egg. More-over, a completely random event deter-mines whether the given chromosome is of male or female origin.

So no two eggs (or sperm) arising in a given individual are ever genetically equiv-alent. No two sexually produced frogs, hav-ing the same two parents, thus will be identical unless they arise by the rare split-ting of an already divided fertilized egg into two separate daughter cells, each of which goes on to develop into a complete embryo. (This is the process by which identical human twins are produced.) In contrast, all the members of a clone pro-duced by mitosis will be identical, except for the occasional mutant cell resulting from rarely occurring somatic gene muta-tions.

The existence of the first clonal frog, the result of the work of the English zoolo-gist John Gurdon, was a very important scientific event. He settled the long-con-troversial biological dilemma of whether the process of cell differentiation in frogs was primarily a cytoplasmic or a nuclear event. During embryological development, the progeny cells which result from the cell divisions commencing after fertiliza-tion become changed (differentiated) into a variety of morphologically and function-ally different cell types. For example, mus-cle cells, nerve cells, and skin cells of a sin-gle individual all have a common ancestor in one fertilized egg.

Most differentiated animal cells, when isolated from contact with other cells, con-tinue to divide and maintain their specific differentiated state. This fact posed the question of the nature of the factors which maintain the specific form of a given differentiated cell. In particular, does differentiation occur through irreversible changes in the nucleus which somehow alter its chromosomal makeup, perhaps by the mutation of specific genes?

Gurdon's clonal frog cleanly settled this point by showing that a nucleus taken from a highly differentiated cell still re-tains its capacity for directing the develop-ment of a completely normal frog. Differ-entiation thus does not involve gene muta-tions. Instead, it must be based upon complicated interactions between the nu-cleus and cytoplasm, interactions which effectively command certain genes to pro-duce the specific gene products needed for a given differentiated state.

The moral issue

But as soon as one such specialized nucleus is removed from its given cyto-plasmic environment, for example by micro-surgical introduction into a new cyto-plasmic environment, the instructions which its genes receive are no longer the same and a new set of genes will go into action. In particular, when the nucleus of a differentiated frog cell is placed inside an enucleated unfertilized egg, it quickly sets in motion the successive steps of em-bryological development leading first to the tadpole stage and finally to an adult frog.

The question of course arises, will this same basic principle hold for the large majority of differentiated cells? Now I suspect most biologists will guess yes. In general, very fundamental phenomena, of which differentiation is one, do not have a different molecular basis from one organ-ism to another. Moreover, it is already clear that differentiation in several plant species does not involve irreversible nu-clear changes. Now it is routinely possible to produce mature plants starting from highly specialized somatic cells of diploid chromosome number. For example, mature carrot plants can be produced from single callus cells that are placed in proper nutri-tional environments. Thus it is highly like-ly that the embryological development of most higher animals, including man, in-volves the creation of countless numbers of somatic nuclei each capable of serving as the complete genetic material for a new organism. This means that, theoretically, all forms of higher animal life may in effect be capable of clonal reproduction.

If true, this situation could have very

116

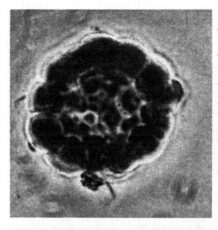

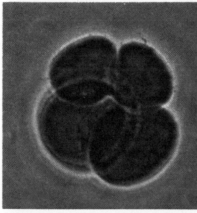

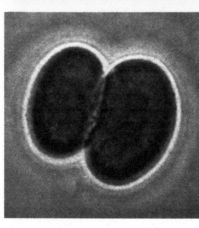

The eggs of a mammalian embryo nourish the 2-cell, 4-cell, and 64-cell stages. At this point, the embryo is implanted on the wall of the uterus, a placenta forms, and the nourishment for later growth comes from the female parent.

The blastula of a metozoan embryo is a hollow, fluid-filled rounded cavity bounded by a single layer of cells.

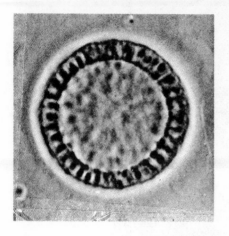

startling consequences as to the nature of human life, a fact soon appreciated by many magazine editors, one of whom commissioned a cover with multiple copies of Ringo Starr, another of whom gave us overblown multiple likenesses of the current sex goddess Raquel Welch. It takes little imagination to perceive that different people will have highly different fantasies; perhaps with some imagining the existence of countless people with the features of Picasso or Frank Sinatra or Walt Frazier or Doris Day. And would monarchs like the Shah of Iran, knowing they might never be able to have a normal male heir, consider the possibility of having a son whose genetic constitution would be identical to their own?

Clearly even more bizarre possibilities can be thought of, and so we might have expected that many biologists, particularly those whose work impinges upon these possibilities, would seriously ponder their implications, and begin a dialogue which would educate the world's citizens and offer suggestions which our legislative bodies might consider in framing national science policies. On the whole, however, this is not at all what has happened.

Though a number of scientific papers devoted to the problem of genetic engineering have casually mentioned that clonal reproduction may someday be with us (the discussion to which I am party) they have been so vague and devoid of meaningful time estimates as to be virtually soporific.

Does this effective silence imply a conspiracy to keep the general public unaware of a potential threat to their basic ways of life? Could it be motivated by fear that the general reaction will be a further damning of all science, thereby decreasing even more the limited money available for pure research? Or does it merely tell us that most scientists do live such an ivory-tower existence that they are capable of thinking rationally only about pure science, dismissing more practical matters as subjects for the lawyers, students, clergy, and politicians to face up to in a real way?

One or both of these possibilities may explain why the occasional scientist has not taken cloning before the public. The main reason, however, I suspect, is that the prospect to most biologists looks too remote and chancy—not worthy of immediate attention when other matters, like nuclear weapon overproliferation and pesticide and auto-exhaust pollutions, present society with immediate threats to its orderly continuation. Though scientists as a group are the most future-oriented of all professions (some investment bankers would probably disagree) there are few of us who concentrate on events unlikely to become reality within the next decade or two.

Cloning of mammals

Behind the general belief that the development of techniques for cloning any mammal, including man, lies far in the future, are fundamental differences between the embryological development of mammals and of amphibians like the frog. These differences reflect the very different environments in which amphibian and mammalian embryos develop. All the frog's embryological development, even in the beginning fertilization stages, occurs in vitro, outside the body of the female parent, generally in the nutrient-poor environment of freshwater lakes and ponds. Thus all the food supply necessary for growth to a developmental stage capable of independent feeding, in the case of a frog to the tadpole stage, must initially be present within the unfertilized egg. As a result, amphibian eggs are not only always relatively large but all their developmental stages are capable of relatively easy experimental investigation.

In great contrast are the eggs of placental, bearing mammals. Their eggs are relatively small, since they have to contain only the nourishment necessary to reach approximately the 64-cell stage. At this point the tiny embryo implants itself on the wall of the uterus, a placenta forms, and all the food molecules necessary for subsequent embryonic growth come from the female parent.

Not only does the small size of such mammalian eggs make experimentation very difficult, but even more important, all the stages of development normally occur within the ovary, oviduct, or uterus. Moreover, there seems to be no real prospect that any mammal can ever be raised totally in vitro. Thus detailed knowledge about the exact steps in the embryogeny of any mammal is much, much less complete than that of amphibians, all of whose development normally occurs in vitro.

The cloning of any mammal thus will be far from a routine task. In particular, the techniques of micromanipulation used to insert nuclei into frog's eggs cannot now be applied to eggs in the mammalian size range. They are likely to be irreversibly damaged by the introduction of a nucleus whose diameter is only some two or three times less than that of the egg itself. And if somehow a trick were ever found to successfully insert a diploid nucleus, the equally challenging task of finding conditions for the in-vitro growth of the modi-

fied egg through to the adult stage would still lie ahead. Thus the clonal production of human beings has seemed to most geneticists an event so unlikely as to not be worth the stirring up of public attention.

This assessment would be correct if the pace of research on human reproductive biology were to continue at the current rate. With a few exceptions, work on the early developmental processes in man has not been seriously pushed either here in the United States or elsewhere. As a result, there exists a scientific lacuna so serious that it deeply disturbs those people who realistically worry about overpopulation problems. They believe that more basic biological knowledge about human reproductive processes would be very helpful in slowing down the fearful rise in the number of human beings.

Consequently, there is already much "population" money available to induce more people to move into the field of reproductive biology, hopefully to learn in great detail the step-by-step processes by which a human egg is ovulated, fertilized, and cleaved, and moves down the oviduct to implant on the uterine wall.

A key ingredient to obtaining this information is the development of methods by which the early embryological stages of mammals can be studied in vitro. For as long as study is restricted to work on intact animals, experimental work, as distinguished from observational analysis, will be virtually impossible.

Nuclear insertion

Most importantly, though unknown even to most biologists, the beginnings of first-rate research on the in-vitro cultivation of mammalian eggs has already occurred. Techniques are in fact available for the isolation of mouse eggs, their fertilization in vitro, and subsequent cultivation under test-tube conditions which permit growth to the 64-cell stage. At this point the embryonic body (called a blastocyst) can be surgically implanted back into the uterus of a living mouse, where it can eventually develop to the stage at which normal birth occurs.

This means that most of the techniques that will be needed to produce a clonal mouse are already available. The only serious obstacles remaining are the development of methods for the removal of the haploid maternal nucleus and the subsequent addition of a diploid adult nucleus. Now there are hints that the enucleation problem will not be serious. For some years it has been known that addition of the mitotic poison colchicine to preovulatory mice leads to abnormal meiotic divisions, which frequently produce nucleus-free eggs. Moreover, very recent work suggests that colchicine in vitro acts similarly. When it is added to unfertilized eggs which have been surgically removed from a living mouse, healthy enucleated eggs are produced.

And furthermore, it looks as if the nuclear insertion process may not be nearly as tricky as it was first thought. This change of opinion comes from the development of very simple methods for the fusing of two cells to yield a single cell containing the genetic compounds of both donor cells. Though the existence of rare examples of cell fusion was first clearly demonstrated in Paris by Barski in 1962, not until 1966 did Henry Harris and John Watkins, working in the pathology department of Oxford University, develop a routine method for easily fusing almost any two desired cells. Their contribution was the introduction of Sendai virus (a close relative of the common flu viruses) killed by ultraviolet light.

In some way not yet understood, adsorption of large numbers of Sendai particles so modifies cell surfaces that when two treated cells touch each other, portions of the opposing cell surfaces effectively dissolve, thereby creating one much larger cell containing two nuclei. Subsequently these nuclei often coalesce, yielding a single nucleus containing all the chromosomes present in both original nuclei.

During the past three years, Christopher Graham, also at Oxford, has been using Sendai virus to fuse mouse eggs with diploid adult mouse cells. The resulting cells still retain the essential features of an egg, because even the relatively small mouse eggs are much larger than most diploid adult cells. While the fused eggs can divide several times, they so far have not yet developed into blastocysts, the stage necessary for successful implantation into the mouse uterus. Conceivably this limitation results from the need to remove the zona pellucida (a normal protective covering) for the Sendai virus fusing trick.

Experimenting with human eggs

Conditions must thus be found either to fuse eggs which retain the zona pellucida or which permit unprotected denuded eggs to develop normally to blastocysts. A reasonable guess is that Graham will succeed, if not this year, most likely within this coming decade. The clonal mammal then will no longer be science fiction.

A likely consequence will be the initiation of similar experiments with a variety of other mammals; first, with easily obtainable laboratory varieties of hamsters, rats, and rabbits, and soon afterwards with economically important domestic animals such as cattle, sheep, and pigs. Though introduction of such methods into animal husbandry might seem at first like economic madness, many veterinarians may suspect otherwise, knowing well the very large prices currently paid for prize animals. Moreover, such research would certainly liven up many agricultural schools, since some of their faculty would jump at something more exciting than the now very routine breeding programs inspired by Mendelian genetics.

So we must expect that, unless somehow strongly discouraged, veterinarians throughout the world someday will attempt the cloning of uniquely valuable domestic animals. One can certainly imagine wealthy racehorse owners wondering whether with a better jockey their prize three-year-old would have been unbeatable. While Nijinsky eventually lost his last two races, might a clonal derivative win every time?

At first consideration, it would seem likely that cloning of many domestic species would have to occur before serious thought would be given to the development of clinical procedures which would make human cloning more than a theoretical possibility. This way of thinking presupposes that the primary purpose for such methodological development need be cloning itself. If this in fact were the objective, the normal and legal objections that would undoubtedly crop up most certainly would effectively prevent the legal granting of the medical facilities needed for extensive in-vitro experiments with human eggs.

If, however, the stated objective is to probe the human reproductive process so that better contraceptive methods can be obtained, the reaction of the general public will be much harder to predict. Though many people will look with horror at any test-tube work with human eggs, others will breathe more easily if something is being done to prevent the world from being crushed by overpopulation. Until several years ago, this latter group was numerically relatively small and without favor in virtually any political circle. Today, however, taboos which would have seemed unbreakable just a decade ago are rapidly being overturned — witness the recent action of the United States Congress in overwhelmingly passing legislation to promote family planning. Even more significant was the action of New York state in making abortions the legal right of any women who desire them.

It thus seems virtually inevitable that for one reason or another the number of people studying all aspects of human embryogeny will greatly increase. Not only will the amount of classical observational analysis increase, but even more impor-

tant, direct experimentation with human eggs most likely will soon be the main preoccupation of a number of intelligent, highly qualified biologists.

Already there exists one such individual, R. G. Edwards, an English reproductive biologist now working in the physiology department of Cambridge University. Originally trained as an embryologist and with some ten years experience in growing mouse embryos in vitro, he focuses his attention on the test-tube growth of human eggs. His original source of material was immature eggs (oocytes) obtained from ovarian tissue that had been surgically excised for reasons completely incidental to his work. From this tissue, Edwards removed the eggs from their surrounding follicles. As such, they were not yet capable of fertilization, since most eggs within human ovaries are present in the dictyotene, a stage just at the beginning of the two meiotic divisions which generate haploid eggs.

But by placing dictyotene-phase eggs in a culture medium similar to that previously worked out for the in-vitro maturation of mouse eggs, the remaining steps of meiosis occur, and some 36 hours later bring the eggs to metaphase II, the stage where normal ovulation occurs. Then when human sperm are added, fertilization occurs, yielding a diploid nucleus capable of dividing several times. However, no fertilized in-vitro matured egg has ever yet developed up to the blastocyst stage. Some factor not yet understood must go wrong during the in-vitro meiosis.

To circumvent this difficulty, Edwards, together with his clinical colleague, P. C. Steptoe of Oldham General Hospital, has devised a simple surgical method for the removal of healthy human eggs after they have completed much of meiosis, but before the ovulation step which releases free eggs from their follicles into the oviduct. Called laparoscopy, it is a relatively minor operation which, while requiring general anesthesia, generally only requires a 24-hour hospital stay.

Help for infertile women

Prior to the operation, a regimen of hormone treatment (gonadotrophins) is given to induce follicle maturation and egg development through the early stages of meiosis. Laparoscopy is then performed, some four hours before ovulation would normally occur. The ovaries so exposed usually contain highly enlarged follicles with thinning walls, through which the desired oocytes can be carefully removed. These procedures have now reached the state where healthy eggs can be obtained from over half the follicles examined.

Such preovulating oocytes are very suitable for subsequent embryological investigations. Fertilization rapidly ensues after the addition of human sperm, and in contrast to those eggs which had undergone meiotic divisions in vitro, these in-vivo matured eggs generally begin normal cleavage divisions. Already many embryos have developed to the eight-cell stage, while a few have become blastocysts, the stage where successful implantation into a human uterus should not be too difficult to achieve. In fact, Edwards and Steptoe hope to accomplish implantation and subsequent growth into a normal baby within this coming year.

The question naturally arises: why should any woman willingly submit to such operations? There is clearly some danger involved every time Steptoe operates. Nonetheless, he and Edwards believe that the risks involved are more than counterbalanced by the fact that their research may develop methods which make their patients able to bear children. All their patients, though having normal menstrual cycles, are infertile, conceivably because many have blocked oviducts which prevent passage of their eggs into the uterus. If so, in-vitro growth of their eggs up to the blastocyst stage may circumvent their infertility, thereby allowing normal childbirth. Moreover, since the sex of a blastocyst is easily determined by chromosomal analysis, such women would be able to decide whether to give birth to a boy or a girl.

Selective cloning

Clearly, if Edwards and Steptoe succeed, their success will be followed up in many other places. The number of such infertile women, while small on a relative percentage basis, is likely to be large on an absolute basis. Within the United States there could conceivably be 100,000 or so women who would like a similar chance to have their own babies.

At the same time we must anticipate strong, if not hysterical, reactions from many quarters. The certainty that the ready availability of this medical technique will open up the possibility of hiring out unrelated women to carry a given baby to term is bound to outrage many people. For there is absolutely no reason why the blastocyst need be implanted in the same woman from whom the preovulatory eggs were obtained. So many women with anatomical complications which prohibit successful childbearing would be strongly tempted to find a suitable surrogate. And it is easy to imagine that many women who just don't want the discomforts of pregnancy would also seek this very different form of motherhood.

Some very hard decisions may soon be upon us. The vague potential for abhorrent misuse should not necessarily weigh more strongly than the unhappiness which thousands of married couples feel when they are unable to have their own children. Different societies are likely to view the matter differently, and it would be surprising if all were to come to the same conclusion. We must, therefore, assume that techniques for the in-vitro manipulation of human eggs are likely to be general medical practice, capable of routine performance in many major nations, within some ten to twenty years.

The situation would then be ripe for extensive efforts, either legal or illegal, at human cloning. No reason, of course, dictates that such experiments need occur. Most of the medical people capable of such experimentation would probably stay totally clear of any step which in any way looked as if its real purpose was to clone. But it would be shortsighted to believe that everyone will instinctively recoil from such purposes. Some people may quite sincerely believe that the world desperately needs many copies of the really exceptional people if we are to fight our way out of the ever-increasing computer-mediated complexity that so frequently makes our individual brains inadequate.

Moreover, given the widespread development of safe clinical procedures for handling human eggs, cloning experiments would not be prohibitively expensive. They need not be restricted to the superpowers — medium-sized, if not minor, countries all now possess the resources needed for eventual success. Furthermore, there need not exist the coercion of a totalitarian state to provide the surrogate mothers. There are already such widespread divergences as to the sacredness of the act of human reproduction that the boring meaninglessness of the lives of many women would be sufficient cause for their willingness to participate in such experimentation, be it legal or illegal. Thus, if the matter proceeds in its current non-directed fashion, a human being — born of clonal reproduction — most likely will appear on the earth within the next twenty to fifty years, and conceivably even sooner, if some nation actively promotes the venture.

The asexually produced child

The reaction of most people to the arrival of this asexually produced child will, I suspect, be one of despair. The nature of the bond between parents and their children, not to mention everyone's values about their individual uniqueness, could be changed beyond recognition if such children became a common occurrence.

Many people, particularly those with strong religious backgrounds, already believe we should *now* deemphasize all forms of research which could lead to circumvention of the normal sexual reproductive processes. If this step were taken, experiments on cell fusion might no longer be supported by federal funds or tax-exempt organizations. Prohibition of such research would most certainly put off the day when diploid nuclei will satisfactorily be inserted into enucleated human eggs. It would be even more effective to take steps quickly to make illegal, or to reaffirm the illegality of, any experimental work with human embryos.

Neither of these prohibitions, however, is likely to be made. In the first place, the cell-fusion technique now offers one of the best avenues for understanding the genetic basis of cancer. Today all over the world cancer cells are being fused with normal cells to pinpoint those specific chromosomes responsible for given forms of cancer.

In addition, fusion techniques are the basis of many genetic efforts to unravel the biochemistry of diseases such as cystic fibrosis or multiple sclerosis. Any attempts to stop such work by using the argument that cloning represents a greater threat than a disease like cancer is likely to be considered irresponsible by virtually anyone able to understand the matter.

Though more people would initially go along with a prohibition against work on human embryos, many may have a change of heart when they ponder the problem which the population explosion poses. The current projections are so horrendous that responsible people are likely to consider the need for more basic embryological facts much more relevant to our self-interest than the not-very-immediate threat of a few clonal men existing some decades ahead. And the potentially militant lobby of infertile couples who see test-tube conception as their only route to the joys of raising children of their own making would carry even more weight. So, scientists like Edwards are likely to get a go-ahead signal even if, almost perversely, the immediate consequences of research supported by "population money" will be the production of even more babies.

Complicating any possible effort at effective legislative guidance is the multiplicity of places where work like Edwards' could occur, thereby making it most unlikely that such manipulations would have the same legal (or illegal) status throughout the world. We must assume that if Edwards and Steptoe produce a really workable method for restoring fertility, large numbers of women will search out those places where it is legal (or possible), just as now they search out places where abortions can easily be obtained.

Thus, all nations formulating policies to handle the implications of in-vitro human embryo experimentation must realize that the problem is essentially an international one. Even if one or more countries stop such research, such action could effectively be neutralized by the response of a neighboring country.

This most disconcerting impotence holds even for the United States. If our congressional representatives, upon learning where the matter now stands, decided they wanted none of it and passed very strict laws against human embryo experimentation, their action would not seriously set back the current scientific and medical momentum which brings us close to the possibility of surrogate mothers, if not of human clonal reproduction. This is because the relevant experiments are being done not in the United States but largely in England. This is partly a matter of chance, but also a consequence of the advanced state of English cell biology. In certain areas it is far more adventurous and imaginative than its American counterpart. There is no American university with the strength in experimental embryology that Oxford possesses.

We must not assume, however, that the important decisions today lie only before the British government. Very soon we must anticipate that a number of biologists and clinicians of other countries, sensing the potential excitement, will move into this area. So even if the current English effort were stifled, similar experimentation could soon begin elsewhere. Thus it appears to me most desirable that as many people as possible be informed about the new ways of human reproduction and their potential consequences, both good and bad.

This is a matter far too important to be left solely in the hands of the scientific and medical communities. The belief that surrogate mothers and clonal babies are inevitable because science always moves forward, an attitude expressed to me recently by a scientific colleague, represents a form of laissez-faire nonsense dismally reminiscent of the creed that American business, if left to itself, will solve everybody's problems.

Just as the success of a corporate body in making money need not advance the human condition, neither does every scientific advance automatically make our lives more "meaningful." No doubt the person whose experimental skill eventually brought forth a clonal baby would be given wide notoriety. But the child, growing up in the knowledge that the world wants another Picasso, would view his creator in a different light.

For the future

Thus I would hope that over the next decade far-reaching discussion will occur, at the informal as well as the formal legislative level, about the many problems which are bound to arise if test-tube conception becomes a common occurrence. On some matters a sufficient international consciousness might develop to make possible some form of international agreement before the cat is totally out of the bag. A blanket declaration of the worldwide illegality of human cloning might be one result of a serious effort to ask the world in which direction it wishes to move.

Admittedly, the vast effort needed

for even the most limited international arrangement will deter those who believe the matter is now of only marginal importance, and that in effect it might be a red herring designed to take our minds off our callous attitudes toward war, poverty, and racial prejudice. But if we do not think about the matter now, the possibility of our having a free choice will one day suddenly be gone.

GENETIC MANIPULATION AND MAN

Darrel S. English

THE SCIENCE OF IMPROVING HUMAN BEINGS by applying the principles of inheritance to obtain a desirable combination of physical characteristics and mental traits is called eugenics. The term was coined by Francis Galton in 1883; literally translated, it means to be "true born" or "well born."

Although most writers on this subject begin by citing the Greeks, the idea of improving the human stock probably goes back even farther. Even though he lacked any knowledge of the laws of heredity, primitive man could see that parents with imperfections often bore children with the same deficiencies. Perhaps the earliest aim was to produce a race of physically perfect men, capable of coping most efficiently with the tremendous hardships they had to endure, including contests with enemies and wild beasts. Paralleling the desire to develop physically was the need to develop intellectually; and cultures may have tended to develop their mental capacities more than their physiques.

Plato advanced the idea of race improvement by methods similar to those of present-day stock breeders. In *The Republic* he proposed that matings between the most nearly perfect men and women be encouraged and that their offspring be raised in a state nursery. Inferior persons should be prevented from reproducing; and if by chance they should have children these should be destroyed. To some extent Plato's eugenic methods were practiced in Sparta; the result was a population of people with fine physiques (Castle, 1925; Fasten, 1935).

In Athens the emphasis was on art, politics, and science. Here, too, people of good background were encouraged to marry among their kind. Did this pay

THE AMERICAN BIOLOGY TEACHER, 1972, Vol. 34, pp. 507-513, 526.

off? Galton (1909) noted that during the 6th to 4th centuries B.C. Athens produced some of the most illustrious men the world has ever known. Whether the decline of Greek society was due to master–slave intermarriage, as some have suggested, or whether the "inferior" classes reproduced more rapidly than the "superior" classes, one can only guess. Possibly the downfall resulted from economic rather than genetic changes.

The eugenic movement appears to have made little headway until the 19th century. **Charles Darwin,** Francis Galton, and Gregor Mendel were among the scientists who kindled the spark of modern eugenics; directly or indirectly, their work stimulated interest in this field. The Darwinian concept of the "survival of the fittest" brought to the fore many inescapable implications. In *The Descent of Man* (1874) Darwin wrote: "It is our natural prejudice and arrogance which made our forefathers declare that they were descended from demi-gods and which leads us to demure to this conclusion." The growing evidence supporting the theory of evolution, together with the refinement of man's ability to influence the evolution of domesticated plants and animals, stimulated work along eugenic lines.

Darwin's cousin Francis Galton, the English anthropometrist and examiner of family records, led the first big surge toward eugenic studies. In his book *Enquiries Into Human Faculty and Its Development* (1883) he coined the term eugenics, defining it as the study of agencies, under social control, that could improve or impair the hereditary qualities of future generations, either physically or mentally (to paraphrase the 2nd ed., 1908). He proposed improvement in human breeding by decreasing the birthrate of unfit persons and increasing the birthrate of fit persons. He made extensive studies on criminality, insanity, blindness, and other human defects. Galton was able to understand the inheritance pattern of some human traits. He recognized the importance of twin studies for human genetics and was aware of the social implications of genetic change in man. He was instrumental in applying more sophisticated statistical methods of solving problems of genetics. (It is interesting to note that Galton, who was unusually gifted and was devoted to the principle that better-qualified people should produce at least their share of children, himself died childless.)

Mendel shed new light on the genetics of man with

the discovery of fundamental principles of inheritance. Thanks in part to his work, scientists learned how to deal with questions of human heredity in a methodical manner.

Eugenic Implementation

Eugenics is often divided into "negative," or preventive, and "positive," or progressive, eugenics. The first is concerned with the elimination of alleles that produce undesirable phenotypes; the second is concerned with furthering the increase of alleles that produce desirable phenotypes or at least with guarding against the decrease of these alleles. In a sense, the two branches of eugenics are identical: to discourage reproduction among people having undesirable traits is, ipso facto, to encourage a comparatively higher rate of reproduction among people who lack these undesirable traits.

In the past, negative eugenics meant sterilization and institutionalization. The results of these measures were insignificant. In many cases, determining what should be considered undesirable was a major problem; furthermore, procedures were not always ethically acceptable to the majority of the people.

More recent methods of negative eugenics include dissuasion from procreation, voluntary sterilization, medically induced abortions, education as to the genetic basis of human traits, and the encouragement of birth-control practices by persons possessing undesirable traits. These efforts are intended to reduce the dysgenic, or deteriorating, effect on society caused by the perpetuation of certain undesirable traits.

Some biochemical defects, which are known to be propagated by a single defective gene, tend to eliminate themselves naturally; many, however, do not. Hemophilia, for example, has severe and often lethal effects generation after generation. In agammaglobulinemia, children are born without the ability to manufacture antibodies. In phenylketonuria (PKU), children are unable to metabolize phenylalanine; they become mentally incompetent if not treated soon after birth. In certain conditions the genetic defect may not be discovered until after reproductive age has been reached and children have already been introduced into the population. Huntington's chorea, with its progressive deterioration of the muscular and nervous systems, does not make itself known until the victim is in his forties. As a result of this dominant lethal gene, approximately 50% of the offspring can expect to succumb to the same fate.

Positive eugenics, aiming at the reproduction of persons of presumedly superior genotypes, has as many problems associated with it as negative eugenics. This concept came into disrepute because of early notions of who was "desirable" and the classification of certain kinds of people as "degenerates." A major setback for eugenicists came during World War II, when Hitler's eugenic movement went to the extreme of trying to achieve a "master race" of "Aryans" at the expense of "non-Aryans." It is understandable why the term eugenics has some very bad connotations in the minds of many people.

Unfortunately, many characteristics that we might consider desirable—high intelligence, good physical health, longevity—are not under the control of a single genetic factor; instead, they arise from a complex of genes that interact in an appropriate environment. H. J. Muller, J. F. Crow, and others have pointed out that such traits as high intelligence and esthetic sensibility have not been selected with any effort in the past. They postulate, however, that if selection for such traits were to be instituted, the general population might respond very rapidly. The means of selection remains the paramount problem. Selection schemes, even though they might be extremely successful, may prove to be completely intolerable.

During the early 1960s, Muller hotly advocated the use of sperm banks in preference to selective-mating programs. These banks would preserve, frozen, the sperm of men of outstanding qualities. This method, called germinal choice, or eutelegenesis, assumes that married women, otherwise barren, might choose to be artificially impregnated with semen from men who had highly desirable traits. This would be possible even though the donor had died several years before conception took place. Full information about the donor would be provided to ensure the best possible combination of genes (Carlson, 1972).

The possibility of such a "preadoption" method has had some acceptance in the United States. It is estimated that 10,000 artificially inseminated conceptions occur in the United States every year (Taylor, 1968). The reason for most births of this kind is that the husband is impotent or possesses some genetic incompatibility, such as the Rh factor, or harbors a known genetic defect, such as hemophilia.

Euphenics

Molecular biologists and medical researchers are developing a series of procedures for the relief of genetic disorders. The field of euphenics is concerned with the improvement of genotypic maladjustments by treatment of genetically defective persons at some time in their life cycle. Today there are many sensitive tests for genetic defects. These enhance the reliability of counseling and decision-making before or during childbearing. As for treatment: in phenylketoneuria, for example, the child is given the "diaper test," which depends on a color change of the urine when ferric chloride is added; or the Guthrie test, which is based on the ability of certain strains of bacteria to grow on substrates containing high levels of phenylalanine. If a problem exists in the metabolism of phenylalanine and the condition is diagnosed early, the infant is put on a diet low in phenylalanine for the first five years of life, and the brain develops normally.

Another example of technologic success in the detection of genetic abnormality is that of a woman who possessed the potential to produce a mongoloid child. The woman had three sisters who suffered from Down's syndrome, or mongolism, and she feared that her own children might have this condition. In 1959 it was discovered that the syndrome is due to an additional (21st) chromosome, which originates by a mistake in cell division just before conception. The woman requested a study of her cells, and her fears were confirmed: she had the extra chromosome. Genetic counselors advised her that she had one chance in three of producing an abnormal child. Several years ago the woman became pregnant. Doctors informed her that a new technique, called amniocentesis, would enable them to determine whether the fetus was aberrant. The method consists in tapping the fluid of the amnionic sac and making chromosome studies of the cells from the fluid. During the 14th week of pregnancy the woman's fears were borne out: she was told her baby was mongoloid. The pregnancy was terminated by therapeutic abortion. Several months later she became pregnant a second time; once again the tests showed Down's syndrome, and the pregnancy was terminated. There was a third pregnancy; and this time the chromosome

130

studies indicated the baby would be normal—and a boy. The woman at last gave birth to what she had wanted for so long: a normal son. A year and a half later, following careful testing, she gave birth to a normal daughter.

Henry Nadler   has used amniocentesis in the diagnosis of high-risk mongolism cases with a high degree of accuracy. This test can be performed between the 12th and 18th weeks of pregnancy. Although there is some danger, the benefits are said to more than justify the risks. Over 35 human enzymatic diseases have been identified by this technique; they include cystic fibrosis, cystinosis, amaurotic idiocy, gout, Gaucher's disease, galactosemia, xeroderma pigmentosum, and diabetes mellitus. Recent advances in the detection of carriers of recessive diseases, such as hemophilia and some forms of muscular dystrophy, are helping to make the job of genetic counseling an easier one (Friedmann, 1971; Nadler and Gerbie, 1971; *Time*, 1971).

Human genetic analysis of single-gene effects by pedigree analysis is still valid and useful. McKusick (1970) listed 1,487 human traits known to be controlled by a single dominant gene, 531 by a recessive gene, and 119 by X-linked genes. With the advent of computerized technology, experiments may now be designed for determining the genetic mechanisms from family-history data, and additional information will be rapidly added to the catalogue of human genetic defects.

From the viewpoint of the population geneticist, the symptoms of the hereditary diseases may have been treated, but the genes remain unchanged and can be passed on to subsequent generations. Therefore the real genetic problem is not solved and, in fact, such medical practices only compound future problems: by preserving defective genotypes and allowing them to reproduce and transmit these genes, we create a population that is more dependent upon surgery, drugs, and similar treatments.

Furthermore, to simply prevent people who possess defective genes from reproducing will not solve the problem. It is estimated that, on average, each person carries four to eight defective genes that in combination with other defective alleles could bereft a child of his normal faculties. In other words, each conception involves some risk of producing a child with a serious abnormality.

The eugenics procedures of the future may be quite different from those of the past. It now appears that the techniques may be at hand to not only "improve" the genetic material by selection but also to correct misinformation within the DNA molecule and thus eliminate the problem.

Genetic engineering, or algeny, is becoming a household word among molecular biologists. Terms such as gene surgery, gene insertion, and gene deletion are beginning to have real meaning for the future of man. The ability to manipulate, in a purposeful manner, the genetic constitution of human beings may come sooner than we anticipate. Taylor (1968) asserted that biologists have reached the critical point of sudden acceleration—the point that physicists reached only a generation ago. It is hoped that biologists and laymen alike will face the problems and potentials of human reproductive engineering in a more realistic and relevant manner than was taken by the atomic physicist and the politicians at the creation of the world's most destructive force (Heim, 1972).

Algeny is a more sophisticated and direct method of curing genetic ills. It is also a permanent cure: in certain instances it might alleviate suffering in future generations. Two approaches seem possible:

1. One might consider the incorporation of a normal gene within the protein coat of some human virus. The cells, after infection with the virus, would supply the defective cell with a corrected copy of the needed information. Although this would only cure the individual during his lifetime and not have any effect on his progeny (unless there might happen to be accidental gene-incorporation), it would be a more direct attack on the genetic problem.

This approach does not appear to be out of the question—as was accidentally demonstrated in a group of laboratory technicians and doctors working with Shope papilloma virus. This virus is capable of inducing tumors in rabbits. Although nonpathogenic in humans, it does infect those who handle it. It appears that one of the viral genes, which is responsible for the synthesis of arginase, is active in human cells. Years after any contact with the laboratory, infected workers were shown to have especially low levels of arginine in their blood as a result of the activity of the viral enzyme (Taylor, 1968).

And now genetic engineers are making their first attempts at using a related means of treating this metabolic disease. Investigators at the Oak Ridge National Laboratories and in Cologne are cooperating in the treatment of two German girls who are suffering from low levels of arginase in their blood. It is hoped that injections of live Shope papilloma virus carrying the enzyme will supplement their low levels of arginase. If this is successful, the girls will be able to produce the needed amino acid, arginine, and so achieve a more nearly normal metabolic balance. Someday it may even be possible to produce viruses artificially to correct specific metabolic deficiencies (Gardner, 1972).

2. A second approach to genetic engineering involves replacement of a mutant gene with a normal one by treating germ cells before fertilization. This curative procedure would be even more direct and would have the advantage of being permanent, because descendants would be normal with respect to the defect in question.

In the microbiologic world two methods of incorporating normal genetic material into a mutant cell have been perfected. Avery, MacCleod, and McCarty (1944) exposed nonvirulent bacteria to purified DNA of a virulent strain of bacteria. To their amazement the nonvirulent bacteria were transformed into virulent forms and were able to transmit the trait to future generations. This process is called transformation.

In 1959 three French workers reported that they had extracted DNA from one strain of ducks, called Campbells, and injected the material into a second strain, Pekins. They had hoped to change the offspring somehow, but to their astonishment the injected Pekin ducks began to take on some of the characteristics of the Campbells. This stimulated a flurry of experiments with these strains of ducks, as well as other animals. But, to the dismay of the researchers, subsequent attempts were a total failure, even with the ducks. Not until 1966 was an actual case of tranformation verified in organisms other than bacteria. A. S. Fox and S. B. Yoons, of the University of Wisconsin, treated one strain of fruit flies with DNA extracts of another strain; some of the resulting offspring developed genetic anomalies that persisted for several generations (Taylor, 1968).

In contrast with transformation is another mode of permanent genetic change effected by using viruses.

This process, called transduction, involves a viral particle that picks up a host gene and, on reinfection in a second host, gives up the genetic material to its new host genome. After incorporation of the newly introduced genome the cell is permanently altered and transmits the acquired trait in the typical genetic manner. The Ukrainian scientist Serge Gerhenson claimed to have transduced a silkworm, using a virus to introduce the foreign DNA. Likewise, there are lines of evidence suggesting that other investigators have been able to transfer drug resistance from one line of mouse-cell cultures to another in this manner (Taylor, 1968).

More recently an exciting experiment was carried out by the molecular biologist Carl Merril and his colleagues   the first successful transplant of bacterial genes into living human cells. Cells from a victim of the genetic disease galactosemia were cultured in vitro. These cells are unable to produce an essential enzyme for the breakdown of the simple sugar galactose. Newborn infants with this defect face malnutrition, mental retardation, and death unless they quickly receive a milk-free diet. Using viruses that had picked up the genes for galactose metabolism from the common intestinal bacterium *Escherichia coli*, the researchers hoped to transmit the gene to human cells in tissue culture. Further investigations showed that the cells had picked up the viruses and that they were producing messenger RNA for the missing enzyme and the enzymes themselves. This clearly implied that the cells were being directed to produce the essential enzyme; and equally exciting was the fact that the enzyme-making capabilities were being transmitted to future generations of cells (Merril, Geier, and Petricciani, 1971). These investigators are now striving to make the same kind of genetic transplants with laboratory animals.

These results are particularly significant because they show that bacterial genes can become biologically active in mammalian cells. Furthermore, they clearly establish the universality of the genetic code. And they could have some important implications for the cure of cancerous conditions produced by "runaway" genes. The field of genetic engineering is indeed wide open. It is enough to make one wonder, though, what new genes a person may pick up from the sneeze of the person sitting next to him!

Clonal Reproduction

Once a highly desirable genotype has been produced, it would be beneficial if more of the same organism could be produced. Asexual reproduction of organisms, including man, is another method that may be used someday to produce desirable genotypes. Cloning is the process of inducing normal somatic cells to repeat the complex step in embryogenesis and eventually produce carbon copies of the original donor organism.

F. C. Steward first showed the feasibility of cloning by taking certain cells from a carrot root and culturing them in coconut milk. Some of these cells formed clumps, which began to differentiate. Transferred to soil, they matured into normal carrot plants. Later experiments have shown that almost any early embryonic cell of the carrot can grow vegetatively (Steward, Mapes, and Smith, 1958).

The possibilities of animal cloning took on reality when J. B. Gurdon, of Oxford University, managed to get the nucleus of an intestinal cell of the South African clawed toad, *Xenopus laevis*, to direct embryogenesis in the enucleated cytoplasm of an unfertilized egg cell. The egg, thus, contained the diploid set of chromosomes and responded by dividing repeatedly. The resulting tadpole was a genetic twin of the toad that had provided the nucleus. By making numerous subclones Gurdon was able to produce many identical copies of the parent toad (Gurdon, 1968). More recently he was able to prepare frogs from cultured cell nuclei (Gurdon, 1970). It would seem there might be no end to the number of copies possible from a single individual. These experiments prove that all the genetic information necessary to produce an organism is encoded in the nucleus of every cell of the organism.

The implications of this technique are numerous. It might be possible to clone a group of Einsteins or Beethovens. Or one might desire a team of astronauts with particular talents and temperaments for a long space voyage. One might be able to clone people with acute psychic awareness, so that extrasensory perception would become commonplace.

A further advantage of this technique has to do with immunologic properties: the members of a clone would be able to accept grafts and tissue or organ transplants without any of the usual repercussions. (The recipient may, however, have some difficulty in convincing one of his clonal twins to give up his heart, lungs, kidneys, or limbs!)

A more practical use of cloning would be its use as a means of tracking extremely deleterious genes during early embryologic development. Someday it may be possible to take the fertilized egg or embryo and culture it in a test-tube. A few cells could be removed and cultured in sufficient numbers for biochemical analysis. If the embryo proved to possess the deleterious genotype, the culture could be terminated; if not, the egg or embryo could be reimplanted in the womb, where development would proceed normally and without further interruption.

Taylor (1968) noted that if vegetative reproduction of human beings is ever achieved, it is most likely to be done by growing a few cells taken from the embryo. The more specialized a cell becomes, the greater the loss of its totipotency. To induce specialized cells, such as nerve, muscle, or brain cells, to become unspecialized once again, appears to be an extremely difficult task.

Of more immediate use to the economy of the world is what one might call a variation on the theme of cloning: the phenomenon of artificial inovulation. All animals seem to be capable of producing many more eggs than they will ever release normally. Injections of the follicle-stimulating hormone (FSH) can induce as many as 40 or more eggs at one time in a cow. This is called superovulation. If the eggs are then artificially fertilized and implanted in a number of competent females, the number of offspring can be multiplied manyfold annually. This process is the converse of artificial insemination, which enables a prize bull to sire more than 50,000 calves a year.

In 1962 two South African ewes gave birth to two lambs whose real parents lived in England. The fertilized eggs had been implanted in the oviduct of a live rabbit, which was flown to South Africa; there the eggs were implanted in the foster ewes. Since then, eggs have been flown to the United States successfully, and transfers between different strains of animals have been accomplished (Taylor, 1968).

Experiments of this kind promise valuable improvements of livestock worldwide. Artificial inovulation also offers hope to those humans afflicted with certain types of sterility. Perhaps the day will come when a woman who would normally be childless will have a prenatally adopted child implanted in her womb, and she will be able to experience all of the emotions and physiologic changes associated with motherhood.

Also related to the ability of man to manipulate the activity of a cell is the phenomenon of regeneration. Mammals have generally lost their totipotency except in specific parts, such as the liver, lymphoid tissue, skin, and bones. The initiatory and regulative factors of regenerative growth are generally unknown. If, however, one could reactivate the genetic events involved in the embryonic organization of cells at the stump of a lost limb, might it not be possible to regenerate the entire structure?

Robert O. Becker, of the Veterans Administration Hospital in Syracuse, N.Y., has said that the possibility of regenerating limbs of mammals, including man, seems nearer. He has successfully stimulated partial regeneration of limbs in rats by the induction of a blastema in response to minute electrical currents applied to the severed area (Becker, 1972). The production of several centimeters of regenerative growth in experimental mammals suggests that higher animals do have regenerative potential if the cells involved can be induced to take on the more primitive state of development. Eventual total regeneration of organs and limbs would make the immunologic complications in transplants a thing of the past.

Conclusion

At one time artificial insemination in humans, sex determination before birth, sex reversal, embryonic determination of genetic aberrations, and organ transplants were looked upon as remote possibilities. Today they are facts; and society has, to some degree, accepted them.

Will man be able to alter his own makeup so profoundly that he will be essentially a new species? Undoubtedly he will strive, with even more enthusiasm, to unlock more of nature's secrets. But some scientists see far greater dangers in man's new knowledge than may be envisioned at first. Seymour Kessler, of Stanford University, has said he "would hate to see manipulation of genes for behavioral ends because as man's environment changes and as man changes his environment, it is important to maintain flexibility" (quoted in Taylor, 1968). One must be cautious about following a path that eliminates variability, because without it we could go the way of the dinosaurs. This is a particular danger if cloning should become popular. Thus, with increased knowledge comes the responsibility to use this new information wisely (Heim, 1972).

Many biologists are concerned over the moral, ethical, and social implications of the new biology. Although predictions of human genetic control are probably premature—many of the techniques are far from perfected—it is only a matter of time before test-tube babies, cloning of humans, and corrective gene-surgery will be possible.

Numerous national and regional commissions have been set up to study some of the problems. Joshua Lederberg, of Stanford University, does not believe that the perfect human being is a proper goal for the molecular geneticist, even if the techniques could be perfected. Nevertheless, there are those who disagree. Who, then, decides what qualities are to be preserved and by whose standards are they to be directed? J. D. Watson, in an article entitled, "Moving toward the Clonal Man: Is This What You Want?" said that he hopes that mankind will thoroughly discuss these issues in the next decade. On the other hand, Glenn T. Seaborg, former director of the Atomic Energy Commission, has stated that he believes decisions should be made by experts without the benefit of public debate. The latter comment reminds us of the warning of the late C. S. Lewis, over a quarter of a century ago: "Man's power over Nature is really the power of some men over other men, with Nature as their instrument." One may agree with Senator Walter F. Mondale, who said: "There may still be time to establish some ground rules." And one may doubt whether we really need a clone of people or the creation of a laboratory full of orphaned test-tube babies at a time when the population explosion is one of man's greatest biologic problems. (For citations of views in this paragraph see Taylor, 1968, and Wallace, 1972.)

And yet . . . genetic engineering does promise to alleviate many of man's ills and sufferings. What a relief from anxiety it would be to know that a genetic disease you might harbor would not be passed on to your child! Can one begin to express the gratitude of a mother as she watches her child, who might have been mentally retarded a few years ago, present the valedictory speech at her high-school graduation? Amniocentesis is likely to show enormous cost benefits in the treatment of Down's syndrome in the near future: 4,000 mongoloid infants are born each year in the U.S.; the lifetime institutional care for each one is approximately $250,000; and therefore, unless women carrying such abnormal fetuses are encouraged to have therapeutic abortions, their care will cost society some $1.75 billion nationally by 1975 (Friedmann, 1971; *Time*, 1971).

One can see that man faces many new legal, moral, and ethical questions. Will he have to consider assigning certain rights to the fetus? At what point does a fetus have legal rights, and to whom does the test-tube baby turn for support and legal inheritance? What will be the nature of the conflict in our society between personal choice and governmental control? These are serious questions, for which today there are no satisfactory answers.

REFERENCES

AVERY, O. T., C. M. MACCLEOD, and M. McCARTY. 1944. Studies on the chemical nature of the substance inducing transformation in pneumococcal types. *Journal of Experimental Medicine* 79: 137-158.

BECKER, R. O. 1972. Stimulation of partial limb regeneration in mammals. *Nature* 235: 109-111.

CARLSON, E. A. 1972. H. J. Muller. *Genetics* 70: 1-30.

DARWIN, C. 1874. *The descent of man*, 2nd ed. D. Appleton Co., New York.

FASTEN, N. 1935. *Principles of genetics and eugenics.* Ginn & Co., New York.

FRIEDMANN, T. 1971. Prenatal diagnosis of genetic diseases. *Scientific American* 225 (5): 34-51.

GALTON, F. 1908. *Inquiries into Human Faculty and Its Development*, 2nd ed. E. P. Dutton Co., New York.

——. 1909. *Essays on eugenics.* Eugenics Education Society, London.

GARDNER, E. J. 1972. *Principles of genetics,* 4th ed. John Wiley & Sons, Inc., New York.

GURDON, J. B. 1968. Transplanted nuclei and cell differentiation. *Scientific American* 219 (6): 24-36.

——. 1970. The transplantation of nuclei from single cultured cells into enucleated frog's eggs. *Journal of Embryological and Experimental Morphology* 24: 227-248.

HEIM, W. 1972. Moral and legal decisions in reproductive and genetic engineering. *American Biology Teacher* 34 (6): 315-318.

McKUSICK, V. A. 1968. *Mendelian inheritance in man: catalogs of autosomal dominants, autosomal recessives and X-linked phenotypes,* 2nd ed. Johns Hopkins Press, Baltimore.

MERRIL, C. R., M. R. GEIER, and J. C. PETRICCIANI. 1971. Bacterial versus gene expression in human cells. *Nature* 233:

398-400.

NADLER, H. L., and A. GERBIE. 1971. Present status of amniocentesis in intrauterine diagnosis of genetic defects. *Obstetrics and Gynecology* 38: 789-799.

STEWARD, F. C., M. O. MAPES, and J. SMITH. 1958. Growth and organized development of cultured cells, I: growth and division of freely suspended cells. *American Journal of Botany* 45: 693-703.

TAYLOR, G. R. 1968. *The biological time bomb.* World Publishing Co., New York.

TIME [magazine]. 1971. Man into superman: the promise and perils of the new genetics. April 19: 33-52.

WALLACE, B., ed. 1972. *Essays in social biology, vol. 2.* Prentice-Hall, Inc., Englewood Cliffs, N.J.

ON RE-DOING MAN *

Kurt Hirschhorn

The past 20 years and, more particularly, the past five years have seen an exponential growth of scientific technology. The chemical structure of the hereditary material and its language have essentially been resolved. Cells can be routinely grown in test tubes by tissue-culture techniques. The exact biochemical mechanisms of many hereditary disorders have been clarified. Computer programs for genetic analysis are in common use. All these advances and many others have inevitably led to discussions and suggestions for the modification of human heredity, both in individuals and in populations: genetic engineering.

One of the principal concerns of the pioneers in the field is the problem of the human genetic load, that is, the frequency of disadvantageous genes in the population. Each of us carries between three and eight genes that, if present in double dose in the offspring of two carriers of identical genes, would lead to severe genetic abnormality, or even to death of the affected individual before or after birth. In view of the rapid medical advances in the treatment of such diseases, it is likely that affected individuals will be able to reproduce more frequently than in the past. Therefore, instead of a loss of genes due to death or sterility of the abnormal, the mutant gene will be transmitted to future generations in a slowly but steadily increasing frequency. This is leading the pessimists to predict that we will become a race of genetic cripples requiring a host of therapeutic crutches. The optimists, on the other hand, have a great faith that the forces of natural evolution will continue to select favorably those individuals who are best adapted to the then current environment. It is important to remember in this context that the "natural" environment necessarily includes man-made medical, technical, and social factors.

Because it appears that at least some of the aspects of evolution and a great deal of genetic planning will be in human and, specifically, scientific hands, it is crucial at this relatively early stage to consider the ethical implications of these proposed maneuvers. Few scientists today doubt the feasibility of genetic engineering, and there is considerable danger that common use of this practice will be upon us before its ethical applications are defined.

A number of different methods have been proposed for the control and modification of human hereditary material. Some of these methods are meant to work on the population level, some on the family level, and others directly on the affected individual. Interest in the alteration of the genetic pool of human populations originated shortly after the time of Mendel and Darwin, in the latter part of the 19th century. The leaders were the English group of

* Based on a paper originally given at a symposium jointly sponsored by Marymount College, N.Y., and Commonweal Magazine.

ANNALS OF THE N.Y. ACADEMY OF SCIENCE, 1970, Vol. 184, pp. 103-110.

141

eugenicists headed by Galton. Eugenics is nothing more than planned breeding. This technique, of course, has been successfully used in the development of hybrid breeds of cattle, corn, and other food products.

Human eugenics can be positive or negative. Positive eugenics is the preferential breeding of so-called superior individuals in order to improve the genetic stock of the human race. The most famous of the many proponents of positive eugenics was the late Nobel Prize winner Herman J. Muller. He suggested that sperm banks be established for a relatively small number of donors, chosen by some appropriate panel, and that this frozen sperm remain in storage until some future panel had decided that the chosen donors truly represented desirable genetic studs. If the decision is favorable, a relatively large number of women would be inseminated with these samples of sperm; proponents of this method hope that a better world would result. The qualifications for such a donor would include high intellectual achievement and a socially desirable personality, qualities assumed to be affected by the genetic make-up of the individual, as well as an absence of obvious genetically determined physical anomalies.

A much more common effort is in the application of negative eugenics. This is defined as the discouragement or the legal prohibition of reproduction by individuals carrying genes leading to disease or disability. This can be achieved by genetic counseling or by sterilization, either voluntary or enforced. There are, however, quite divergent opinions as to which genetic traits are to be considered sufficiently disadvantageous to warrant the application of negative eugenics.

A diametrically opposite solution is that of euthenics, which is a modification of the environment in such a way as to allow the genetically abnormal individual to develop normally and to live a relatively normal life. Euthenics can be applied both medically and socially. The prescription of glasses for nearsighted individuals is an example of medical euthenics. Special schools for the deaf, a great proportion of whom are genetically abnormal, is an example of social euthenics. The humanitarianism of such efforts is obvious, but it is exactly these types of activities that have led to the concern of the pessimists, who assume that evolution has selected for the best of possible variations in man and that further accumulations of genes considered abnormal can only lead to decline.

One of the most talked-about advances for the future is the possibility of altering an individual's genetic complement. Since we are well on the way to understanding the genetic code, as well as to deciphering it, it is suggested that we can alter it. This code is written in a language of 64 letters, each one determined by a special arrangement of three out of four possible nucleotide bases. A chain of these bases is called deoxyribonucleic acid, or DNA, and makes up the genetic material of the chromosomes. If the altered letter responsible for an abnormal gene can be located and the appropriate nucleotide base substituted, the corrected message would again produce its normal product, which would be either a structurally or enzymologically functional protein.

Another method of providing a proper gene, or code word, to an individual having a defect has been suggested from an analysis of viral behavior in bacteria. It has long been known that certain types of viruses can carry genetic information from one bacterium to another or instruct a bacterium carrying it to produce what is essentially a viral product. Viruses are functional only when they live in a host cell. They use the host's genetic machinery to translate their own genetic codes. Viruses living parasitically in human cells can cause such diseases as poliomyelitis and have been implicated in the causation of tumors. Other viruses have been shown to live in cells and to be reproduced along with the cells without causing damage either to the cell or to the organism. If such a harmless virus either produces a protein that will serve the function of one lacking in an affected individual, or if it can be made to carry the genetic material required for such functions into the cells of the affected individual, it could permanently cure the disease without additional therapy. If carried on to the next generation, it could even prevent the inheritance of the disease.

Transplanting Nuclei

An even more radical approach has been outlined by Lederberg. It has become possible to transplant whole nuclei, the structures that carry the DNA, from one cell to another. It has become easy to grow cells from various tissues of any individual in tissue culture. Such tissue cultures can be examined for a variety of genetic markers and thereby screened for evidence of new mutations. Lederberg suggests that it would be possible to use nuclei from such cells derived from known human individuals, again with favorable genetic traits, for the asexual human reproduction of replicas of the individuals whose nuclei are being used. For example, a nucleus from a cell of the chosen individual could be transplanted into a human egg whose own nucleus has been removed. This egg, implanted in a womb, could then divide just like a normal fertilized egg, to produce an individual genetically identical to the one whose nucleus was used. One of the proposed advantages of such a method would be that, as in positive eugenics, one could choose the traits that appear to be favorable, and do so with greater efficiency by eliminating the somewhat randomly chosen female parent necessary for the sperm bank approach. Another advantage is that one can mimic what has developed in plants as a system for the preservation of genetic stability over limited periods of time. Many plants reproduce intermittently by such vegetative or parthenogenetic mechanisms, always followed by periods of sexual reproduction for the purpose of elimination of disadvantageous mutants and increase in variability.

Another possibility derives from two other technological advances. Tissue typing, similar to blood typing, has made it possible to transplant cells, tissues, and organs from one individual to another with reasonably long-term success. During the past few years, scientists have also succeeded in producing cell hybrids containing some of the genetic material from each of two cell types, either from two different species or from two different individuals of the same species. Very recently, Weiss and Green at New York University have succeeded in hybridizing normal human culture cells with cells from a long-established mouse tissue-culture line. Different products from such fusions contain varying numbers of human chromosomes and, therefore, varying amounts of human genes. If such hybrids can be produced which carry primarily that genetic information which is lacking or abnormal in an affected individual, transplantation of these cultured cells into the individual may produce a correction of his defect.

These are the proposed methods. It is now fair to consider the question of feasibility. Feasibility must be considered not only from a technical point of view; of equal importance is the effect of each of these methods on the evolution of the human population and the effect of evolution on the efficacy of the method. In general, it can be stated that most of the proposed methods either now or in the not too distant future will be technically possible. We are, therefore, not dealing with hypothesis in science fiction but with scientific reality. Let us consider each of the propositions independently.

Positive eugenics by means of artificial insemination from sperm banks has been practiced successfully in cattle for years. Artificial insemination in man is an everyday occurrence. But what are some of its effects? There is now ample evidence in many species, including man, of the advantages for the population of individual genetic variation, mainly in terms of flexibility of adaptation to a changing environment. Changes in environment can produce drastic effects on some individuals, but a population that contains many genetic variations of that set of genes affected by the particular environmental change will contain numerous individuals who can adapt. There is also good evidence that individuals who carry two different forms of the same gene, that is, are heterozygous, appear to have an advantage. This is true even if the gene in double dose—that is, in the homozygous state—produces a severe disease. For example, individuals homozygous for the gene coding for sickle cell hemoglobin invariably develop sickle cell anemia, which is generally fatal before the reproductive years. Heterozygotes for the gene are, however, protected more than normals from the effects of the most malignant form of malaria. It has been shown that women who carry the gene in single dose have a higher fertility in malarial areas than do normals. This effect is well known to agricultural geneticists and is referred to as hybrid vigor. Fertilization of many women by sperm from few men will have an adverse effect on both of these advantages of genetic variability because the population will tend to be more and more alike in its genetic characteristics. Also, selection for a few genetically advantageous factors will carry with it selection for a host of other genes present in the same individuals, genes whose effects are unknown when present in high numbers in the population. Therefore, the interaction between positive eugenics and evolution makes this method not feasible on its own.

Abnormal Offspring

Negative eugenics is, of course, currently practiced by most human geneticists. It is possible to detect carriers of many genes that, when inherited from both parents, will produce abnormal offspring. Parents, both of whom carry such a gene, can be told that they have a one-in-four chance of producing an abnormal child. Individuals who carry chromosomal translocations are informed that they have a high risk of producing offspring with congenital malformations and mental retardation. But how far can one carry out such a program? Some states have laws prescribing the sterilization of individuals mentally retarded to a certain degree. These laws are frequently based on false information regarding the heredity of the conditions. The marriage of people with reduced intelligence is forbidden in some localities, again without adequate genetic information. While the effects of negative eugenics may be

144

quite desirable in individual families with a high risk of known hereditary diseases, it is important to examine its effects on the general population.

These effects must be looked at individually for conditions determined by genes that express themselves in a single dose (dominant) or in double dose (recessive) and those which are due to an interaction of many genes (polygenic inheritance). With a few exceptions, dominant diseases are rare and interfere severely with reproductive ability. They are generally maintained in the population by new mutations. Therefore, there is either no need or essentially no need for discouraging these individuals from reproduction. Any discouragement, if applicable, will be useful only within that family but not have any significance for the general population. One possible exception is the severe neurological disorder, Huntington's chorea, which does not express itself until most of the patient's children are already born. In such a situation it may be useful to advise the child of an affected individual that he has a 50% chance of developing the disease and a 25% chance of any of his children's being affected. Negative eugenics in such a case would at least keep the gene frequency at the level usually maintained by new mutations.

The story is quite different for recessive conditions. Although detection of the clinically normal carriers of these genes is currently possible only for a few diseases, the techniques are rapidly developing whereby many of these conditions can be diagnosed even if the gene is present only in single dose and will not cause the disease. Again, with any particular married couple it would be possible to advise them that they are both carriers of the gene and that any child of theirs would have a 25% chance of being affected. However, any attempt to decrease the gene frequency of these common genetic disorders in the population by prevention of fertility of all carriers would be doomed to failure. First, we all carry between three and eight of these genes in single doses. Secondly, for many of these conditions, the frequency in the population of carriers is about one in 50 or even greater. Prevention of fertility for even one of these disorders would stop a sizable proportion of the population from reproducing and for all of these disorders would prevent the entire human population from having any children. Reduction in fertility of a sizable proportion of the population would also prevent the passing on to future generations of a great number of favorable genes and would, therefore, interfere with the selective aspects of evolution, which can function only to improve the population within a changing environment by selecting from a gene pool containing enormous variability. It has now been shown that in fact no two individuals, with the exception of identical twins, are likely to be genetically and biochemically identical, thereby allowing the greatest possible adaptation to changing environment and the most efficient selection of the fittest.

The most complex problem is that of negative eugenics for traits determined by polygenic inheritance. Characteristics inherited in this manner include many measurements that are distributed over a wide range throughout the population, such as height, birth weight, and intelligence. The last of these can serve as a good example of the problems encountered. Severe mental retardation in a child is not infrequently associated with perfectly normal intelligence or in some cases even superior intelligence in the parents. These cases can, *a priori*, be assumed to be due to the homozygous state in the child of a gene leading to mental retardation, the parents representing heterozygous carriers. On the other hand, borderline mental retardation shows a

145

high association with subnormal intelligence in other family members. This type of deficiency can be assumed to be due to polygenic factors, more of the pertinent genes in these families being of the variety that tends to lower intelligence. However, among the offspring of these families there is also a high proportion of individuals with normal intelligence and a sprinkling of individuals with superior intelligence.

All of these comments are made with the realization that our current measurements of intelligence are very crude and cannot be compared between different population groups. It is estimated that, on the whole, people with superior intelligence have fewer offspring than do those of average or somewhat below average intelligence. If people of normal intelligence were restricted to producing only two offspring and people of reduced intelligence were by negative eugenics prevented from having any offspring at all, the result, as has been calculated by the British geneticist Lionel Penrose, would be a gradual shift downward in the mean intelligence level of the population. This is due to the lack of replacement of intellectually superior individuals from offspring of the majority of the population, that is, those not superior in intellect.

Current Possibilities

It can be seen, therefore, that neither positive nor negative eugenics can ever significantly improve the gene pool of the population and simultaneously allow for adequate evolutionary improvement of the human race. The only useful aspect of negative eugenics is in individual counseling of specific families in order to prevent some of the births of abnormal individuals. One recent advance in this sphere has important implications from both a genetic and a social point of view. It is now possible to diagnose genetic and chromosomal abnormalities in an unborn child by obtaining cells from the amniotic fluid in which the child lives in the mother. Although the future may bring further advances, allowing one to start treatment on the unborn child and to produce a functionally normal infant, the only currently possible solution is restricted to termination of particular pregnancies by therapeutic abortion. This is, of course, applied negative eugenics in its most extreme form.

Euthenics, the alteration of the environment to allow aberrant individuals to develop normally and to lead a normal life, is currently being employed. Medical examples include special diets for children with a variety of inborn errors of metabolism who would, in the absence of such diets, either die or grow up mentally retarded. Such action, of course, requires very early diagnosis of these diseases, and programs are currently in effect to routinely examine newborns for such defects. Other examples include the treatment of diabetics with insulin and the provision of special devices for children with skeletal deformities. Social measures are of extreme importance in this regard. As has many times been pointed out by Dobzhansky, it is useless to plan for any type of genetic improvement if we do not provide an environment within which an individual can best use his strong qualities and obtain support for his weak qualities. One need only mention the availability of an environment conducive to artistic endeavor for Toulouse-Lautrec, who was deformed by an inherited disease.

146

The feasibility of alteration of an individual's genes by direct chemical change of his DNA is technically an enormously difficult task. Even if it became possible to do this, the chance of error would be enormous. Such an error, of course, would have the diametrically opposite effect of that desired; in other words, the individual would become even more abnormal. The introduction of corrective genetic material by viruses or transplantation or appropriately hybridized cells is technically predictable and, since it would be performed only in a single affected individual, would have no direct effect on the population. If it became widespread, it could, like euthenics, increase the frequency in the population of so-called abnormal genes, but if this treatment became a routine phenomenon, it would not develop into an evolutionarily disadvantageous situation. It must also be constantly kept in mind that medical advances are occurring at a much more rapid rate than any conceivable deterioration of the genetic endowment of man. It is, therefore, very likely that such corrective procedures will become commonplace long before there is a noticeable increase in the load of disadvantageous genes in the population.

The growing of human beings from cultured cells, while again possibly feasible, would interfere with the action of evolutionary forces. There would be an increase, just as with positive eugenics, of a number of individuals who would be alike in their genetic complement, with no opportunity for the high degree of genetic recombination that occurs during the formation of sperm and eggs and which is evident in the resultant progeny. This would diminish the adaptability of the population to changes in the environment and, if these genetic replicas were later permitted to return to sexual reproduction, would lead to a marked increase in homozygosity for a number of genes with the disadvantages pointed out before.

Who Will Be the Judges?

We see, therefore, that many of the proposed techniques are feasible although not necessarily practical in producing their desired results. We may now ask the question, which of these are ethical from a humanistic point of view? Both positive and negative eugenics as applied to populations presume a judgment of what is genetically good and what is bad. Who will be the judges, and where will the line be between good and bad? We have had at least one example of a sad experience with eugenics in Nazi Germany. This alone can serve as a lesson on the impossibility of separating science and politics. The most difficult decisions will come in defining the borderline cases. Will we breed against tallness because space requirements become more critical? Will we breed against nearsightedness because people with glasses may not make good astronauts? Will we forbid intellectually inferior individuals from procreating despite their proved ability to produce a number of superior individuals? Or should we, rather, provide an adequate environment for the offspring of such individuals to realize their full genetic potential?

C. C. Li, in his presidential address to the American Society of Human Genetics in 1960, pointed out the real fallacy in eugenic arguments. Man has continuously improved his environment to allow so-called inferior individuals to survive and reproduce. The movement into the cave and the

147

putting on of clothes protected the individual unable to survive the stress of the elements. Should we then consider that we have reached the peak of man's progress, largely determined by environmental improvements designed to increase fertility and longevity, and that any future improvements designed to permit anyone to live a normal life will only lead to deterioration? Nineteenth-century scientists, including such eminent biologists as Galton, firmly believed that this peak was reached in their time. This obviously fallacious reasoning must not allow a lapse in ethical considerations on the part of the individual and by humanity as a whole, just to placate the genetic pessimists.

The tired axiom of democracy that all men are created equal must not be considered from the geneticist's point of view, since genetically all men are created unequal. Equality must be defined purely and simply as equality of opportunity to do what one is best equipped to do. When we achieve this, the forces of natural evolution will choose those individuals best adapted to this egalitarian environment. No matter how we change the genetic make-up of individuals, we cannot do away with natural selection. We must always remember that natural selection is determined by a combination of truly natural events and the artificial modifications that we are introducing into our environment in an exponentially increasing number.

With these points in mind, we can try to decide what, in all of these methods, is both feasible and ethical. I believe that the only logical conclusion is that all maneuvers of genetic engineering must be judged for each individual and in each case must take into primary consideration the rights of the individual. This is by definition impossible in any attempt at positive eugenics. Negative eugenics in the form of intelligent genetic counseling is the answer for some. Our currently unreasonable attitude toward practicing negative eugenics by means of intelligent selection for therapeutic abortion must be changed. Basic to such a change is a more accurate definition of a living human being. Such restricted uses of negative eugenics will prevent individual tragedies. Correction of unprevented genetic disease, or that due to new mutation, by introduction of new genetic material may be one answer for the future; but until such a new world becomes universally feasible, we must on the whole restrict ourselves to environmental manipulations from the points of view both of allowing affected individuals to live normally and of permitting each individual to realize his full genetic potential. There is no question that genetic engineering will come about. But both the scientists directly involved and, perhaps more important, the political and social leaders of our civilization must exercise utmost caution in order to prevent genetic, evolutionary, and social tragedies.

The Issues: Social, Ethical, and Legal

This section of the book contains a potpourri of articles reflecting diverse positions on the social, ethical, and legal issues related to genetic counseling and engineering. Four editorials from scientific journals introduce some of the issues. Richard Lewontin in an editorial from BioScience confronts us with the interesting idea that there are ethical issues that cry for our attention which are of far greater significance than those associated with genetic counseling and engineering. He notes that "...the activities of science exploit or neglect whole classes of people, taking advantage of their powerlessness."

The two editorials from Science present a somewhat more conservative picture of the prospects of genetic engineering than one gets from much popular writing. They point to some of the dangers associated with arousing public hope, fear and concern over these prospects. Certainly, as Abelson notes, one of the most serious and undesirable consequences of this fear and concern would be the placing of harmful restrictions on scientific research. In fact, what is actually needed, as Fox and Littlefield note, is more public support for basic research in human biology. In the fourth editorial Littlefield emphasizes that the average citizen currently has both the desire and the right to know about his own genetic constitution. Littlefield cites the need for better education of the general public with respect to human biology, including human genetics.

Medical science currently has at its command techniques for large scale screening for a number of human genetic defects. For example, not only can individuals having sickle cell anemia be detected, but also the heterozygous carriers of the "recessive" gene for sickling. Since asymptomatic heterozygotes for such genetic diseases can be identified, they can be advised in a genetic counseling program of their potential for transmitting the deleterious gene to their offspring. All of this poses ethical and social issues that are discussed in a brief article on bioethics from Laboratory Management. The article includes guidelines for operating a genetic screening program which were originally published in the New England Journal of Medicine. These guidelines are designed to protect the individuals and families screened and to lessen the risk of misuse of the information obtained.

At the present time the technology for such procedures as cloning and genetic surgery have not been developed to the point where they can be used in the genetic engineering of

man. We have a period of grace in which to clarify our
thinking with respect to the ethical and social problems
these procedures raise. But the time for considering the
problems created by genetic counseling is now! Genetic
screening procedures followed by genetic counseling are being
used now, are certainly going to be more widely used in the
future, and are perhaps even going to become mandatory by
legal action in some cases. This is an issue that will not
"go away" - we must face it now. A portion of Sorenson's
paper, "Social Aspects of Applied Human Genetics", continues
the discussion of genetic counseling and its implications.
The role of the counselor and the reproductive decisions that
must be made after genetic counseling has been given are among
the topics Sorenson includes.

In an article from the American Biology Teacher Werner
Heim differentiates between the roles of the biologist, the
technologist, and the social scientist and humanist with respect to the study and application of human genetic principles. Heim suggests that the social scientist and humanist
have the responsibility of assisting us all in deciding
"...whether, how, and under what conditions the work of the
other two [biologist and technologist] ought to be applied."
This does not absolve the biologist from all responsibility
for the social and ethical implications of his discoveries.
It does suggest, however, that the biologist lacks the expertise of the social scientist, lawyer, philosopher or theologian in considering the ethical and social significance of
the application of his discoveries to human genetic improvement.

U. S. Senator from California, John V. Tunney and his
lawyer associate Meldon E. Levine look at the moral, ethical
and legal implications of genetic engineering. Tunney and
Levine stress the importance of public awareness and education and emphasize that consideration must be given to these
problems now rather than at some future date when"...we
might face irreversible trends not only in genetics but also
in political freedom."

As a sequel to the Tunney-Levine article, Saturday Review provides us with a brief look at one attempt to control
human reproduction through in vitro fertilization followed by
the implantation of the young embryo in a receptive uterus.
The possible significance of such an experiment and the objections of Nobel laureates Watson and Perutz are noted.

In the final selection in this part of the book Walter G.
Peter III reports on an international conference which dealt

with ethical perspectives in the use of genetic knowledge. The article includes a number of thought-provoking observations, not the least of which is the fact "...that the uses to which genetic knowledge will be put will not be decided by those who best understand the 'new biology', but by those who understand it the least, the public...."

SCIENCE AND ETHICS

RICHARD C. LEWONTIN

As science, and especially biological science, has gained ever-increasing power to alter our species' relationship to nature and the relationship of one human being to another, there has been a growing preoccupation on the part of scientists with what are conceived of as the "ethical" problems that arise from that power. This preoccupation, however, has been confined almost exclusively to individual ethical decisions, related to problems of individual choice and freedom. Thus, we are told that if we can diagnose genetical disorders *in utero*, a serious ethical problem arises in the decision to deprive a possibly debilitated foetus of its right to life. In a more futuristic vein, we are asked to ponder the weighty ethical problem that arises if we could manipulate an individual's genes at birth or before, since we would be making a decision about a person's biological nature and the biological nature of future generations. Or, again, we are reminded of the serious ethical issues that face the clinical investigator who must decide whether to use an untested and unsure treatment on a seriously ill patient.

What is seldom realized is that the preoccupation with individual moral issues such as these, however serious they may be, is the result of a class bias peculiar to scientists, academics, and other middle class persons. For such privileged persons for whom personal freedom and choice are taken for granted, these individual interferences with liberty and destiny seem fraught with significance. But they are utterly blind to the fact that large groups of human beings are victims, by *socially determined necessity,* of scientific decisions and research priorities. Why is it that most blood donors in America are either the social outcasts of the cities, or persons who cannot afford to pay for the blood needed by their hospitalized relatives, and so must pay with their own life's blood? Why is it that clinical experimentation is carried out on prisoners, conscientious objectors, the indigent institutionalized, and others who can be induced to trade their bodies for the promise of freedom or a few precious and scarce privileges? Why is it that South Africa's most famous heart transplant had as its recipient a white well-to-do dentist and as its donor a black? How much does a heart transplant cost, anyway, and can someone on welfare get one? What is the meaning of the fact that large sums are newly put into "genetical engineering" and other elegant and sophisticated genetical research, but that black Americans are not routinely screened at birth for the inherited deficiency in the activity of the enzyme glucose-6-phosphate dehydrogenase, which is a potentially lethal disorder present in 14% of black men and 2% of black women? Who pays for the medical research funded by the NIH, and who are the chief beneficiaries?

BIOSCIENCE, 1971, Vol. 21, p. 799.

The questions go on and on, but they seem not to concern our moral philosophers. While the biologist indulges his moral fancies in the ethics of amniocentesis or the problem of whether an extra Y chromosome may foreshadow a life of crime, he turns his back on the fact that the activities of science exploit or neglect whole classes of people, taking advantage of their powerlessness.

The chief and overwhelming ethical and moral issue facing us is that the organization of our political economy guarantees that a large fraction of human beings will be the victims of the omissions and commissions of science because they lack the material wealth and the social power to control their own lives. If biologists are really concerned with the ethics of their activities, they should turn their attention to this social question, and they must have the courage to follow wherever that examination leads.

ANXIETY ABOUT GENETIC ENGINEERING

Philip H. Abelson

Science and technology have provided society with innumerable options and the ability to influence evolution. Optimists see in this a great future, with enhancement of the quality of life and of the dignity of mankind. Pessimists see quite a different picture, and at present they appear to be the more numerous and influential. The average citizen, when he thinks about it, is uncomfortable with the necessity of judging complex issues without adequate facts or background. He also feels relatively powerless to affect the outcome.

In spite of the widespread feeling of ineffectiveness, some people have had very great influence and, collectively, the electorate is having profound effects. Public concern about misuse of technology is leading to measures designed to cope with such present abuses as pollution. Technology can be geared to ameliorate part of the disagreeable conditions, and constructive steps are now being taken.

Some of the difficulties created by science and technology are not so close to solution, particularly in biology and medicine. Advances in these fields have led to great benefits and also to puzzling problems, including some for which our present ethical concepts do not prepare us. More technology alone is hardly likely to provide satisfactory answers to the population explosion. Especially disturbing are aspects of the measures taken to prolong life in the very sick and very old. Death of a loved one was bad enough when it was in the hands of God; now it is often a much more distressing experience. Since every individual must participate in birth and death, he cannot escape some thought about the associated problems that science has created; in general, he is not grateful for the necessity to face such issues.

During the last several years, the public has repeatedly been warned that science is creating additional problems through raising the possibility of test tube babies and "genetic engineering." The response of the public has been negative, with some calling for a halt to research in molecular biology. In truth, the dire predictions of the potentialities of new science have outrun the accomplishments, and the predictors have assumed that society will exercise negatively the options that are provided. Speculation about test tube babies is based on a modest accomplishment—that is, fertilizing a human egg in vitro and keeping it alive for a week or so.

SCIENCE, 1971, Vol. 173, Issue No. 3994. Copyright 1971 by the American Association for the Advancement of Science.

For many years, biologists have been fertilizing eggs of countless species in vitro. Talk of genetic engineering received impetus from the isolation of an operon, a specific piece of DNA. This accomplishment is about as meaningful as the isolation of other components of the living system. Biochemists are experts at taking life apart, and they can reassemble some subsystems. The total system, however, is orders of magnitude more complex than anything they have put together. Even if biochemists achieve a capacity for genetic engineering, it is unlikely that their tools will match the tools that are already available. For example, artificial insemination is widely used to improve livestock. If some future ruling clique decided to engage in human genetic improvement, they would be more likely to adopt this technique and to employ their own semen than to use material concocted in the laboratory.

Talk of the dire social implications of laboratory-related genetic engineering is premature and unrealistic. It disturbs the public unnecessarily and could lead to harmful restrictions on all scientific research.

RESERVATIONS CONCERNING GENE THERAPY

Maurice S. Fox
John W. Littlefield

The attention recently given the prospects of gene therapy requires a realistic appraisal of the potential as well as a sober consideration of the liabilities of this therapeutic approach.

There is no doubt that the development of techniques for transfer of genes and chromosomes in laboratory studies of mammalian cells will provide a powerful research tool toward comprehension of both normal and abnormal cellular processes and will ultimately provide a rationale for the treatment of many human diseases. Gene therapy, however, involves direct application of this technology to individuals suffering from genetic disease. Possibilities under discussion include: introduction of DNA or of chromosomes either directly or by somatic cell fusion; transfer of genetic material from one host to another by virus-like particles containing DNA of the host cell; infection with active or inactive virus containing genes that can determine some particular biochemical function; or infection with a viral nucleic acid to which some cellular gene has been coupled.

Although the number of newborns suffering from disorders that can be described as genetic is very large, only a small fraction of these disorders would even in principle be amenable to intervention by any of these techniques. Neither genetically dominant disorders, nor multigenic traits, nor disorders resulting from extra chromosomes could be alleviated. The major remaining class is that of the recessive "inborn errors of metabolism." These occur with a collective frequency of about 1 per 1000 individuals and include, conservatively, between 100 and 1000 different disorders. Gene therapy would be likely to involve the isolation of somatic cells from a diseased individual, the alteration of their genetic endowment in vitro, and their replacement in the individual. For example, it seems unlikely that sickle-cell anemia would be relieved if a few percent of the blood-forming cells were replaced by cells capable of producing normal hemoglobin, or that the consequences of phenylketonuria would be relieved by the presence of a few somatic cells capable of converting phenylalanine to tyrosine. On the whole, it does not seem probable that more than a small fraction of the inborn errors could be helped by these techniques, and, with new developments

SCIENCE, 1971, Vol. 173, Issue No. 3993. Copyright 1971 by the American Association for the Advancement of Science.

in the understanding of the immune response, these disorders will probably be treated more easily and effectively by tissue transplantation or some sort of enzyme therapy.

Furthermore, there are certainly hazards, both known and unknown, that accompany the presently conceived strategies. Many of the procedures are likely to be mutagenic, and who can guess how many dominant effects, visible only in the whole individual, might appear? Most of the viruses under consideration as vectors are tumor-producing. Even the fractionated virus-like particles containing cell DNA are certain to include some particles containing viral DNA. Damaging alterations of regulatory processes and even uncontrolled tumor-like growth could easily be the consequences of introducing additional chromosomes or a host of viral genes.

The promises offered by the proponents of gene therapy largely ignore its limitations and hazards. To mislead the public in this regard risks another period of disappointment and reaction. We are still primarily in a descriptive phase in our understanding of human genetics, with little, if any, idea of how to intervene safely at any level. Let us not do to ourselves what we have done to our environment. Let us now seek public support for research toward a better understanding of normal and abnormal human biology, rather than promise quick glamorous cures.

GENETIC SCREENING

John W. Littlefield, M.D.

THE comprehensive guidelines for genetic screening developed by the Institute of Society, Ethics and the Life Sciences, which appear elsewhere in this issue of the *Journal*, should certainly be required reading for anyone setting up such a program. But being primarily cautionary, they do not convey the enthusiasm many of us have for this sort of screening or its potential scope.

Definition of the genetic makeup of an individual for the sake of his own health and that of his offspring is the goal of medical genetics. To illustrate how far we are from this, it suffices to note that nowadays the "at-risk" couple is almost always detected by the appearance of affected offspring. The number of cases that can subsequently be prevented in such families is relatively small. If at-risk parents were detected before reproduction, they could choose to take the risk or not, or to use adoption, artificial insemination, or prenatal diagnosis when available. Also, from the public-health point of view, such a total program would be necessary to decrease the frequency of various deleterious genes.[1]

Genetic screening is in fact not new at all, but has been concerned to date mainly with detection of the individual whose *own* health is threatened. Thus, the screening of at-risk families (10 to 50 per cent risk of a second case) is well accepted — e.g., the siblings of a child with a recessive disorder such as phenylketonuria or the relatives of a patient with a dominant disorder such as multiple polyposis of the colon. Among less frequent conditions (1 to 10 per cent risk), we could also be screening the siblings of children with urinary-tract malformations, for example, at least for symptoms if not by laboratory tests. Already, many people are screened at some time for diabetes mellitus. Some testing of "high-risk" groups for inherited drug sensitivities has started, such as the testing of male newborns at the Boston City Hospital for glucose-6-phosphate dehydrogenase deficiency. Much more of this sort of

THE NEW ENGLAND JOURNAL OF MEDICINE, 1972, Vol. 286, pp. 1155-1156.

screening should be done,[2] and doubtless will be as pharmacogenetic technics and education improve. Screening for the hyperlipoproteinemias that produce premature vascular disease, especially Type II, is also just around the corner. Regarding rare conditions (frequency of less than 1 per cent), the screening of newborn infants for inborn errors of amino acid metabolism by the Massachusetts Department of Public Health has been notably comprehensive, instructive and well received, such problems as have been encountered concerning technics and cost-effectiveness[3] rather than ethical questions. If all newborn infants were also screened routinely for major chromosome abnormalities (as is now done at the Boston Hospital for Women), the yield of cases would be over 10 times greater, but the absence of treatment for such abnormalities, the present uncertainty about the importance of some of them, and the cost involved discourage such a program now. This situation may change with automation, and if balanced translocations and other rearrangements are indeed as frequent (1.5 per 1000) and important as appears to be the case.[4] Ultimately, all metabolic and chromosomal screening might best be done in utero, if we can be assured that the method is quite harmless, and when congenital malformations and dominant disorders such as osteogenesis imperfecta and tuberous sclerosis can be diagnosed in the fetus in this way. Genetic examination of the fetus would not only characterize heterozygous and homozygous states but also detect new mutations or chromosome disorders and circumvent problems due to illegitimacy at a point in time when abortion was an option or intrauterine treatment might be indicated.

Now attention is shifting to the detection of those whose genes threaten the health of their future offspring. Again, we have been doing such screening already when the frequency of the genetic risk was high (10 to 50 per cent), as in the screening for Rh-incompatible couples. Likewise, for serious X-linked diseases in which the carrier state can be diagnosed with reasonable certainty, many geneticists believe that examination and counseling should be offered to the adult female relatives of an affected male, especially since prenatal diagnosis has become available. This obligation may even take precedence over the traditional extent and confidentiality of the patient-doctor relation, if one believes that people have the right to know about the risks that they run, whether infectious, toxic, or genetic, and to be protected from these risks if they wish. The same obligation occurs upon detection of a familial chromosome rearrangement.

Primarily, the Institute's guidelines are directed toward the many programs now starting to detect in high-risk ethnic groups those (1 to 10 per cent frequency) whose genes place their offspring in jeopardy when combined with similar genes in their mates. Such programs aim at the sickle trait and other hemoglobinopathies in persons of African descent, and the genes of Tay–Sachs disease and certain other sphingolipidoses in persons of Ashkenazic Jewish origin. Public funds for the former are becoming available, and the latter programs await private support and some final improvements in technic.[5] Screening for the thalassemia gene in other populations could begin anytime, and that for the cystic-fibrosis gene in Caucasians, it is hoped, will be available soon. Economically, however, screening does not seem appropriate now for rarer recessive genes (less than 1 per cent frequency), such as for alpha$_1$-antitrypsin deficiency and Fanconi's anemia, which in the homozygous state respectively cause emphysema and predispose to neoplasia. Indeed, even in the heterozygous state these genes, like that for sickle trait, may not be entirely innocuous[6,7]; the total burden of disease in such heterozygotes might be surprisingly large, justifying their inclusion in screening programs.

But the intention of the "new" genetic screening to examine and inform essentially healthy people about their genes, and, it is hoped, educate them in the process, raises new psychosocial and ethical problems even if the screening is "voluntary." Do people really want to know so much about their genes? My own affirmative answer derives from counseling experiences, a belief that people have the right to know, and the feeling that daily hammering about DNA in the schools and newspapers has made everyone less scared about genes, and more curious. The idea that we all carry silently a few very bad ones has been around quite a while, but few people realize that there are many other genes that differ somewhat from one person to the next and contribute importantly to human individuality. What we should do now is put DNA back in perspective, and from elementary school on teach *human* biology, including anatomy, psychology, genetics, etc. We might then understand and accept our weaknesses and those of others a bit earlier, practice a little more preventive medicine, and cope more wisely with some of the enormous biosocial problems around us.

REFERENCES

1. Motulsky A, Fraser GR, Felsenstein J: Public health and long-term genetic implications of intrauterine diagnosis and selective abortion. Intrauterine Diagnosis (Birth Defects Original Article Series Vol 7, No 5). Edited by D Bergsma. New York, The National Foundation, 1971, pp 22-32

2. Riccardi VM, Hecht F: Military ethics and G-6-PD deficiency. N Engl J Med 286:947-948, 1972

3. Shih VE, Levy HL, Karolkewicz V, et al: Galactosemia screening of newborns in Massachusetts. N Engl J Med 284:753-757, 1971

4. Lubs H: Neonatal cytogenetic surveys, Perspectives in Cytogenetics: The next decade. Edited by SW Wright, BF Crandall, L Boyer. Springfield, Illinois, Charles C Thomas, 1972, pp 297-304

5. Brady RO, Johnson WG, Uhlendorf BW: Identification of heterozygous carriers of lipid storage diseases: current status and clinical applications. Am J Med 51:423-431, 1971

6. Talamo RC: The α_1-antitrypsin in man. J Allergy Clin Immunol 48:240-250, 1971

7. Swift M: Fanconi's anaemia in the genetics of neoplasia. Nature (Lond) 230:370-373, 1971

BIOETHICS:
A MOUNTING DILEMMA FOR GENETICISTS

The explosion of genetic knowledge and the development of sophisticated diagnostic procedures have sparked a growing concern over the ethical application of genetic technologies. Each step forward seems loaded with the double-edge potential for "ambush or opportunity," as Cal Tech geneticist Dr. Richard Sinsheimer puts it.

The prospects are promising or frightening, depending upon one's perspective. Phillip Handler, past president of the National Academy of Sciences, has suggested that the gene pool—humanity's genetic legacy to future generations— might be purged through selective abortion, that "some people are dispensable." Other prominent scientists at least admit the necessity of discussing the matter. Dr. Margery Shaw, director of the Medical Genetics Center at Houston, Texas, asks, "Do our genes belong to us, or do they belong to future generations?" But Dr. Marc Lappe, Associate for Biological Sciences at the Institute of Society, Ethics and the Life Sciences (Hastings Center) in Hastings-on-the-Hudson, N.Y., counters, "To maintain that we are adding to the public good by conscientious genetic guidance and screening may be fallacious and ethically dangerous. The genetic counselor's obligation never should extend beyond the family within his purview."

The only common denominator among all concerned may be recognition of the immediate need for discussion. Conferences of scientists, theologians, philosophers and lawyers are convening regularly to lay the format for a "bioethics." Last December a bill passed the senate to fund a two-year commission that would evaluate the need for a permanent National Advisory Commission on Health Science and Society to monitor the progress of technology with respect to its effects on society. More recently, an offshoot of the World Health Organization, the Council for International Organization of Medical Sciences, passed a resolution

LABORATORY MANAGEMENT, 1972, Vol. 10, No. 11, pp. 34-37.

recommending that an international committee be established to consider the moral and social issues precipitated by recent breakthroughs in the life sciences.

Are basic rights at risk? Will freedom of choice be supplanted by coercive measures? Does man lurk in ambush for man? Or, as Dr. Sinsheimer believes, is man about to play a pivotal role in his own evolution?

Fostering new dilemmas

The Hastings Center was founded in 1970 in direct response to questions such as these. Its contention is that new technologies foster moral, ethical and legal dilemmas which must be met in an "orderly, sustained and creative way" and then translated into viable public and legal policies. Because these issues are multi-disciplinary in scope, the Hastings Center brings together members of the scientific, legal and philosophical communities under five research groups— Ethics and Population Limitation, Death and Dying, Behavior Control, Medical Ethics, and Genetic Counseling and Genetic Engineering.

Though significant breakthroughs in genetic engineering and gene therapy seem imminent, the Genetics task force is currently focusing its attention upon a more contemporary and subtle form of manipulation—genetic counseling. Members of that research group contend that precedents now being established with regard to mass screening programs and counseling will be valuable in assessing future advances.

As director of the Genetics task force, experimental pathologist Dr. Marc Lappe is concerned with investigating what he views as the widening chasm between genetic technologies and the ethics of their application. He grapples daily with the apparently insoluble conflict between scientific fact and social values.

Illustrative of this conflict is the case of individuals afflicted with sickle cell anemia, or those who carry the sickle cell trait in their genes. Although no specific therapy for that disease is now available, several potential remedies—cyanate therapy, for instance—appear likely in the near future. The advent of an effective unsickling agent will reverse the destinies of those now fated to die in childhood. They could lead normal lives and have children of their own—all with the sickle cell trait. Thus, although the disease would be controlled, the trait would be perpetuated.

How is genetic counseling to proceed in this situation, given the dilemma of individual cure at the "expense" of propagating the sickle gene in the population at large? Will selective abortion of homozygotes in each genration become compulsory by law? Will mass sterilization become a reality?

163

And who is to be screened for the disease, the ethnic group which displays the greatest propensity toward it, or all others who carry the potential in their genes?

These and many more considerations blur the role of the genetic counselor almost beyond definition. Certainly there are no simple answers at hand. At a recent international conference, "Ethical Issues in Genetic Counseling and the Use of Genetic Knowledge," jointly sponsored by the Hastings Center and the NIH's John E. Fogarty International Center, Daniel Callahan, director of the Hastings Center, summed up his report by noting that a consensus of opinion was never reached on many issues.

Eugenics or situation ethics?

Three markedly disparate points of view seem to dictate the *modus operandi* of genetic counselors. Although few appear to subscribe exclusively to one operational philosophy, most counselors seem to gravitate toward one of two poles. One extreme represents the eugenic approach which seeks to maintain or improve the genetic potential of the human species. The other is a "situation ethics" in which the counselor considers only the family immediately within his care. Balancing these two poles from the center is the third approach—the traditional role of the genetic counselor as a neutral educator, who merely presents the facts which will permit his patient to exercise her freedom of choice.

In practice, however, specific cases rarely, if ever, dovetail neatly into these descriptive niches. The leading question which the counselor finds himself more frequently asking is not only "What do I do with the facts?" but may be "What are the facts?" In a case study for the Hastings Center, Robert M. Veacth, Associate for Medical Ethics, gives an example of this quandry. The case history involves a woman, unexpectedly pregnant at age 41, at high risk for the birth of a mongoloid child. Knowing this, the woman opted for an amniocentesis, deciding that she would abort if mongolism was indicated. A chromosomal analysis of the fetal cells showed no extra 21st chromosome, and the possibility of translocation mongolism was therefore discounted. The genetic counselor was ostensibly relieved of his contract—to ascertain the likelihood of a mongoloid child—with the woman.

However, the fetus' sex chromosome composition was found to be an abnormal XYY, suggesting a vaguely defined condition known as "supermale." Some research findings hint that persons with an XYY sex composition may be inclined toward violent acts. Other research of the XYY syndrome dispells any such notion. Veacth labels this problem "a condition of doubt."

Explains Veacth, "The counselor has to act; he has to decide something. It is the type of delimma which is becoming more common in the medical-scientific professions. Some very human problems arise for which we have only a bit of scientific evidence, a promising theory, or perhaps only a wild hypothesis. Some day there may be more evidence, but a decision is demanded today."

Meticulously, Veacth aligns the arguments and counter-arguments for and against "telling." On one end of the balance rests the moral imperative, "When there is real doubt, wait." On the other side lies the tenet, "Proceed unless there is good reason not to." And so a tenuous equilibrium is maintained. Nor does that mainstay of the Hippocratic Oath—Do no Harm—offer much solice. Harm to whom? The parents, or the fetus, or both? And, more importantly, what constitutes harm in such an ill-defined situation? If the counselor elects not to reveal the information which he stumbled across, isn't he relieving the parents of potentially unnecessary anxiety and suffering? To extend the thesis of that argument further to include a consideration of the fetus, might not the most serious harm of all—death through abortion—be averted?

Conversely, equally satisfying arguments lead to the conclusion that the parents should be advised of their son's extra Y chromosome. Says Veacth, "If they later learn about their child's extra Y chromosome and hear or read about its implications, they may be subject to much greater suffering and anxiety than if they are told now. Their anxiety in this case would be compounded by loss of confidence in their counselor."

Point and counter-point entice the counselor deeper into an apparently insoluable labyrinth. Each principle seems to have an equally attractive and valid obverse. And looming over the entire dilemma is the inescapable fact that some decision is demanded. Obviously, no clear-cut solution presents itself. Ultimately, each counselor's own ethical system of reasoning must break the deadlock. Regardless of the choice, one overwhelming fact seems clear: Limited technologies which have yielded science only fragmented scraps of knowledge may have banished, at least for the present, the counselor's neutral role. The neutral educator is obliged only to present the facts. But what are the facts?

Once the facts have been discerned, they can be interpreted and presented to the patient according to one of two basic biases—eugenic, which dictates a societal responsibility, or situation ethics in which the immediate and only concern is the patient and her family. This of course discounts the role of neutral educator, but there is little evidence that in practice that stance is ever assumed. On the basis of a sociological study of the decision-making process in genetic counseling, J.R. Sorenson of Princeton University believes that regardless of how avidly neutrality is espoused, counselors do in fact dictate decisions to their patients, if only through the ways in which they present the facts. The patient-physician interaction is not immune from degrees of subjectivity.

Responsible to whom?

To whom then is the genetic counselor ultimately responsible, society or the individual? Dr. Marc Lappe contends that the immediate impact of their decisions must take precedence over any future impact, supposed or real, and thus obligate the counselor only to the family within his purview. "Basically," says Lappe, "I think that genetic counselors may be misguided if they feel that their ethical obligation is in *any way* to future generation. My basic precept is that in genetic counseling all we may *ever* be able to do is to *minimize* the bad associated with human reproduction."

Dr. Lappe dismisses the function of the counselor as an agent of eugenic policy by mounting a convincing attack against the purported goal of negative eugenics, defined as the *elimination* of the genetically "unfit." According to his argument, the western ethic of the "greatest good for the greatest number" ignores many of the realities of human genetics. Central to his reasoning is the contention that genetic "improvement" with respect to a complete elimination of homozygous recessive individuals for a specific disease state, is an unattainable end "since the rate of reintroduction of the gene plus the continued breeding of all heterozygotes—often with reproductive compensation—far outstrips the rate at which the gene can be eliminated through abortion."

Even more pointedly, Lappe argues that the counselor with an eye on the future cannot accurately and compassionately assess the immediate problems which his patient is likely to encounter. Often the future obscures the present; more often, a supposed future "societal good" is at odds with the counselor's immediate mandate to do no harm.

"The best that one can do in the role of genetic counselor (as that role is currently construed) in the face of the inevitability of genetic defects is to minimize the impact—not necessarily the occurrence—of these defects," argues Lappe. "This impact is measured by the amount of human suffering coincident to the defect."

166

> **BIOETHICS GIVES RISE TO**
> **GENETIC SCREENING GUIDELINES**
>
> In response to legislators' queries and to focus atten-
> tion on problems of stigmatization, confidentiality and
> breaches of the right of privacy and freedom of choice
> in genetic screening programs, the Hastings Center
> recently proposed the following guidelines: (1) the
> need for well planned, attainable program objectives,
> (2) involvement of all communities in designing pro-
> gram objectives, (3) provision for equal access by all,
> (4) adequate and accurate testing procedures, (5)
> absence of compulsory measures, (6) procedures for
> obtaining informed consent, (7) safeguards for pro-
> tecting test subjects, (8) open access of communities
> and individuals to program policies, (9) provision of
> follow-up counseling services, (10) education as to the
> relation of screening to realizeable or potential thera-
> pies, (11) protection of family and individual privacy.
> (For the full statement see the New England Journal
> of Medicine, 286:21. May 25, 1972.)

Nevertheless, there are those who advocate a eugenic policy. As procedures such as amniocentesis, chromosome analysis, and biochemical assays become increasingly available to more accurately ascertain the presence of delet- erious genes, abortion is viewed by some as the means to wipe out genetic disease much the same as the vaccine conquered polio.

Most eugenic rationales, in fact, seem to derive from comparisons of genetically-related diseases with the management of other disease states—tuberculosis and syph- ilis, for example. Communicable infectious diseases, runs the argument, are transmitted horizontally from person to person. In the case of contagious diseases, a physician is obliged to extend his concern beyond the individual to safeguard against societal damage. Laws exist which stipulate the isolation of persons with certain infectious diseases.

The eugenecist similarly argues that genetic diseases are contagious, the difference being that they are transmitted vertically from parent to offspring. Carriers of genetic disease can be isolated very effectively via sterilization to prevent spread of the disease which is latent in their genes. As long as freedom of choice prevails, some agree that this argument may carry some weight. In escalating numbers, men are opting for vasectomy, though more for reasons of birth control than to prevent a deleterious genetic legacy. The question which concerns others is: Will freedom of choice become societal responsibility enforced through coercive measures?

Bruce Hilton, Associate for publications at the Hastings Center, writes in a recent issue of *The Hastings Center Report:* "Amniocentesis and abortion bestow a great and welcome freedom from the burden of rearing children with certain diseases. But how soom will that freedom become a responsibility—in the eyes of society—to abort certain fetuses whether we want to or not? The sympathy toward parents 'afflicted' with a mongoloid child will change in the day when we know they *chose* to have that child. And when will public opinion solidify into law? When will genetic disease, whose impact is potentially far greater than that of VD, become a legally *reportable* disease like VD or a quarantinable disease (by prohibition against marriage between carriers)? What will happen when civil liberties increasingly conflict with public health priorities?"

Taken individually, the incremental steps toward change seem innocent enough. Massachusetts passes a law requiring sickle cell screening for children as a prelude to elementary school admission. A fluorescence test on fetal cells in California informs a mother in Kansas that her child will be male. A scientist at the Oak Ridge National Laboratory in Tennessee freezes mouse embryos and implants them into surrogate mothers at some later date. Even more exotic happenings—human clones and man-animal hybrids—are clearly visible on the horizon.

The public embraces each of man's wonderful achievements lavished upon them, and is shielded from the implications of genetic technology by the sheer complexity of its nature. Reports one obstetrician, "Women today ask for an amniocentesis as if it were no more complicated than an appendectomy."

But in retrospect, future man will assess each incremental change in the *aggregate* and will have little doubt that his predecessors were responsible for some awesome accomplishments. The question is: Will he bless or damn his ancestors for what they have "achieved."

168

SOCIAL ASPECTS OF APPLIED HUMAN GENETICS

By James R. Sorenson

Social Aspects of Applied Genetics

Recent advances in science and medicine are increasing man's control of the quantity and quality of human populations. The development of highly effective and relatively efficient birth control techniques permits expanded regulation of population size and growth. In addition, advances in medical genetics are making possible growing intervention and manipulation of the genetic quality of human populations.

Progress in medical genetics can alter man's role in evolution. Man is no longer limited to passive acceptance of all inherited characteristics but is rapidly expanding his technological capacity to include the active treatment, selection, and elimination of many individual genetic attributes. These developments pose complex questions of a moral, ethical, political, psychological, or economic nature. For instance, what genetic attributes or constitutions are desirable? Who is to decide? Should genetic anomalies be reduced in a population? Who shall say how or when? As with most technological developments, knowledge that permits increasing intervention and control of the genetic quality of life is accumulating more rapidly than is man's ability to apply this knowledge wisely.

Some of the technological developments that permit control of the quality of human life have not precipitated serious problems. In

SOCIAL SCIENCE FRONTIERS, 1971, No. 3, pp. 1-14. Copyright 1971 by Russell Sage Foundation.

most Western societies medical research and practice have achieved near complete control of many of the major infectious diseases that have plagued mankind for centuries; such control has met little resistance. A case in point is the development of polio vaccine and the subsequent virtual elimination of poliomyelitis.

The success of such medical advances was dependent on several things, especially prevailing values. For example, the implementation of programs to control infectious disease required a value system in which disease was interpreted as a natural event. If parents felt shame or guilt for the infectious diseases suffered by their children, attempts to treat the disorders and to develop curative methods would have been obstructed. Also, such a value system had to provide for the approval of man's active intervention to control disease. The doctor in Western society has not only received approval for his intervention in disease control but also has achieved a great deal of esteem and prestige from the society he has served.

The idea of treatment and intervention in man's genetic health is not universally accepted today, by either the general public or the medical profession. This is due in part to existing values and beliefs. Parents often experience guilt and shame for genetic disorders in their offspring (Lynch, et al., 1965). More important perhaps, many people including some medical personnel, believe that medical procedures are not capable of correcting or treating any genetic disorders, or that doctors and parents should not attempt genetic intervention (Lynch, 1969).

Through the ages man has interpreted the significance of genetic anomalies in many ways. Sometimes, anomalies were interpreted as the favor of the gods, and at other times they were considered to be portents of divine wrath (Reisman and Matheny, 1969). Today where these beliefs still linger, they limit the use of genetic knowledge by the public and by medical practitioners. Counterbalancing these beliefs and values is a rapidly increasing technological capacity to treat and to select certain genetic conditions. The eventual role of applied human genetics in medical practice and society will reflect the complex intertwining of existing values and beliefs with increasing technical capacity. Neither factor alone will predict man's orientation toward an intervention in his genetic future.

170

One of the earliest applications of genetic knowledge in medical practice was genetic counseling, also referred to as "inheritance counseling." This is a form of medical service through which people gain information about their genetic constitution or that of their children. Genetic counseling has had a short history. For much of this history doctors were limited to informing families that they had a disease or disorder that "ran in the family." Such medical counsel was of little practical use to families. With the advent of Mendelian genetics and much later the linkage of selected diseases with specific modes of inheritance, genetic counseling became more precise. Doctors could give parents with an identifiable genetic disorder statements of the risk that any child they had would exhibit the genetic defect. For example, a couple might have had a child who died from cystic fibrosis, a lethal genetic disease of childhood. A counselor making this diagnosis would inform the couple that they faced a risk of approximately 1 in 4 of having another child with cystic fibrosis (Carter, 1969). The couple could then use this information to make a decision about future reproductive behavior.

Genetic counseling, as practiced until recently, was limited primarily to the calculation and delivery of such risk statements. Prospective parents could evaluate the risk in a given case and then act on their decision. If they chose to take a risk, they had to live with the child whether he was normal or not.

Several recent advances are changing the nature of genetic counseling. The development of amniocentesis, a procedure by which fetal chromosomes are analyzed for abnormalities, permits intra-uterine detection of several genetic and most major chromosomal defects (*Fogarty International Center Proceedings,* 1970). Using this technique, parents may choose to abort a fetus with an identifiable defect and to have only healthy children. For example, Down's Syndrome, more commonly known as mongolism, a congenital moderate-to-severe form of mental retardation due to specific chromosomal abnormalities, can be detected in utero with almost total accuracy and little apparent risk to the mother or fetus (*Fogarty International Center Proceedings,* 1970). With this procedure the mother need no longer carry such a fetus to term, if the parents so decide and if their doctor and the state concur. The risk of having such a child can thus be removed. Recent developments also permit the detection of an increasing number of genetic defects in persons who carry a genetic anomaly but do not clinically exhibit the disease. For example, carriers of sickle-cell anemia, a form of anemia due to an abnormal type of hemoglobin in red blood cells, limited almost entirely to black populations, can be identified (Carter, 1969). This procedure permits afflicted couples to be advised of their condition prior to having children. Thus, genetic counseling can be given before the birth of any afflicted children, rather than after, as in the past.

In addition to an expanded technical base, many other factors have increased the potential application of genetics in medical practice. First, the role genetic endowment plays is being delineated in more and more diseases. It apparently plays a minor role in the etiology of many diseases, such as cancer and heart trouble; a moderate role in others, such as diabetes; and a major, if not determinate, role in a third class of disease, such as cystic fibrosis, sickle-cell anemia, and Down's Syndrome (Carter, 1969). Until now, genetic counseling has been limited primarily to diseases falling in the third class, but it can be used to advise individuals who have diseases of the first two classes. In these situations clients can be informed of the risk of occurrence of the disease and can be advised to take precautionary measures, such as dietary and environmental management, to reduce their risk of disease. In addition, increasing research efforts and discoveries in behavioral genetics, while not yet applicable, indicate that genetic counseling

172

might eventually involve the role that inheritance plays in the appearance of selected aspects of psychological functioning and possibly some aspects of social behavior (Lindzey, et al., 1971).

Of the diseases in the population that are major threats as health problems, those with a genetic base are constituting an increasingly larger proportion (Carter, 1969). In controlling infectious diseases man has found that imperfections or peculiarities in his own genetic constitution are becoming health threats. This factor will certainly operate to increase the demand for early and extensive appilcation of medical genetics to human populations by both professional medical groups and the public.

Changes in public attitudes and practices are increasing public demand also for more extensive programs in medical genetics. With the increasingly widespread acceptance of population control, it seems reasonable that parental concern for the health of their smaller families will increase. This could include concern for the genetic health of their children. In addition, the acceptance of birth control practices reflects new values and changed attitudes toward reproductive behavior. Not only can reproduction be planned, but with the new developments mentioned above the health of many children who risked disease can now be reasonably assured.

Finally, there may be considerable economic advantages in developing mass screening programs of people likely to pass on specific genetic diseases to their offspring (Scriver, 1970). Therapeutic abortion of fetuses afflicted with Down's Syndrome or other severely disabling diseases detectable in utero can save families much sorrow, and save both the state and the families from a severe financial burden (Danes, 1970). The development of intra-uterine detection techniques makes such screening programs feasible. The changing moral and legal climate surrounding abortion suggests that the number of therapeutic abortions will increase in the near future.

Increased application of genetics in medicine will not necessarily be automatic, however. The use of genetic knowledge in medical practice has given rise to many controversial issues in the recent past and continues to do so. How large a risk should prospective parents take? Should parents make the decision? Should abortion of a fetus be permitted on the grounds that it will be abnormal? Should carriers of severe genetic defects be forbidden to marry, or to have children? These questions are in every respect—ethically, morally, politically, emotionally, legally—difficult to answer. More questions will arise as knowledge of human heredity advances.

The increasing intervention in and control of genetic disease and the selection or avoidance of certain genetic constitutions have fostered debates within and between many different groups in society. Medical professionals, life scientists, lawyers, ministers, biologists, and philosophers are engaged in discussions concerning the proper use of such knowledge. A prerequisite for intelligent discussion of these issues is information about how genetic knowledge is used today. Which doctors give genetic counseling? Which parents seek counseling? What types of reproductive decisions do people make when faced with genetic disorders? Very little information is available on these topics. If man is to understand and to employ the potential good inherent in medical genetic advances, he must begin by determining by whom and how these advances are used. The social sciences, by providing such information, can make significant contributions to discussions on the use of genetic knowledge.

Human Genetics in Medical Practice:
A Review and Analysis

Because the application of human genetics in medical practice is relatively new, few studies explore the psychological and sociological aspects of genetics in medicine. The existing literature is both meager and scattered.

There are numerous medical publications concerning the clinical nature of genetic counseling, which is the major use of medical genetics today (Sorenson, 1971a). This literature includes discussions of problems that genetic counselors face, and the types of services they provide. A small amount of literature reports the general public's knowledge of specific genetic disorders. In addition, there are a few studies of the reproductive decisions made by couples after genetic counseling. There are no detailed studies of the distribution in the United States of medical genetic facilities or their services. Finally, little mention is made of the economic aspects of medical genetics.

Most of the available information has been reported by individual medical geneticists or counselors and, being based on their personal experiences, is of limited use in generalizing about medical genetics. There is thus considerable need for extensive social science research on applied human genetics.

Public Knowledge and Use of Medical Genetics

Extensive sociological literature suggests that knowledge about certain kinds of infectious diseases and action taken to alleviate those diseases is inversely related to social class (Feldman, 1966; Mechanic, 1968). The assumption that knowledge about various genetic disorders would vary according to social class is substantiated by existing literature.

In a national survey Feldman (1966) found considerable variation in public knowledge about various diseases. For example, whereas nearly 70 percent of his sample knew at least one correct symptom of polio, only 48 percent could correctly name a symptom of diabetes, a genetically based metabolic disorder (Feldman, 1966). More specifically, knowledge about diabetes was positively associated with education and income, established indicators of social class. The experiences of many genetic counselors support this observation. Their practice suggests that in the past, and to a large extent today, there has been little utilization of medical genetic facilities by the poor (Juberg, 1966).

The fact that historically the majority of people receiving genetic counseling have been of the middle or upper social classes is a reflection of several factors. First, the middle class, and apparently their doctors as well, have been more informed than other classes about the availability of medical genetic services. Second, since the use of genetic counseling has in the past depended largely on self-referral, the willingness of the middle class and the reluctance of the lower classes to seek needed medical assistance have been important (Mechanic, 1968). Economic constraints, while of some importances in shaping the use of medical services, are not always of primary concern (Myers and Schaffer, 1954). Until recently, most genetic counseling was free, with minimal charges for necessary laboratory work. The factors determining the distribution of services appear to have been essentially special knowledge of the availability of such services, plus sets of values about and attitudes toward the procreative process that legitimated intervention and control.

The social distribution of health information is itself determined in part by the social sources of that information, that is, the process of knowledge dissemination. Feldman (1966) suggests that for diseases such as cancer and polio, essentially nongenetic diseases, a primary source of information is the mass media. His research, as well as several additional studies (Koos, 1954; Deshaies, 1962), suggests variation across social classes, however, in the relative

176

importance of various information sources for knowledge of infectious disease. Generally, the mass media are basic information sources for the more educated segments of the population, while personal contacts serve as primary information sources for the less educated segments (Feldman, 1966). Doctors play a minor role in educating the public. Their role is apparently limited to informing the patient about the particular disease he has (Feldman, 1966).

Public knowledge about genetic disorders does not appear to be disseminated in the same way as knowledge about nongenetic disorders, especially for the more educated segments of the population. Feldman reported that the most important source of information about diabetes was an afflicted relative or close friend. This suggests that general knowledge about genetically based diseases depends more upon interpersonal channels of communication than does knowledge about infectious diseases, and that individuals are likely to be acquainted with such diseases only if a family member or close friend is afflicted.

The social distribution of knowledge about genetic disease reflects the fact that genetic considerations are of limited interest to most people. Concern for these disorders is generally confined to those afflicted, or to those in their reproductive years who have family or friends suffering from genetic disorders. Experiences of counselors indicate that genetics is of minor concern for most people when they select marriage partners. Requests for premarital counseling constitute a small segment of a counselor's time (Reed, 1963). In those cases the most frequent questions concern consanguinity, not the existence of a genetic problem in one partner or the other. The most frequent requests for counseling come from the parents of a genetically defective child (Reed, 1963). It is at this point that the parents become concerned about genetic health and seek counseling to avoid further defective children or simply to relieve anxiety. Counselors are also often contacted for advice by state and local agencies such as state institutions and adoption agencies. Occasionally, counselors are contacted for assistance by the courts in a paternity suit (Reed, 1963).

Taken together, these observations suggest that in the past and at present medical genetic services are not used by all who could benefit from them, nor are they used at the most opportune time. While medical genetics is inherently a preventive form of medicine, it seems to be used in response to crisis situations—after genetic problems occur, rather than before.

The Nature and Scope of Genetic Counseling

There are many forms of genetic counseling. The people who practice genetic counseling have various educational backgrounds: some are research scientists, others medical doctors (Sorenson, 1971b). Genetic counseling is not yet a medical specialty. Because most physicians lack training in human genetics, counseling is usually limited to those who have specialized knowledge about human genetics, although they may or may not be medical doctors.

In the late 1940's and early 1950's there were perhaps 10 or 12 genetic counselors in the United States (Reed, 1963). Today there are about 200 counseling units (Lynch and Bergsma, 1971). With mounting discoveries and an increasing awareness of the need for counseling, this impressive growth should continue in the near future.

The most common task facing the genetic counselor is the determination of the risk of recurrence of a genetic defect in a family (Reed, 1963). Supported by Mendelian genetics, the notions of dominant and recessive autosomal or x-linked modes of inheritance, the counselor can often provide specific information about recurrence risks (Lynch, 1969). Although many genetic disorders may follow the classical inheritance modes of Mendelian ratios, many do not. In these cases the counselor must rely upon empirical statistics of risk. These statistics have been derived from studying the incidence of a disease in many families (Neel, 1958). From these observatons the counselor can provide the clients with approximations of the recurrence risk of a specific disease. When the counselor is equipped with neither Mendelian ratios nor empirical estimates of risk, he can inform his clients only of his ignorance.

In addition to statistics of risk the counselor today can employ a battery of new techniques that permit increased detection of genetic and chromosomal abnormalities in utero. These techniques include amniocentesis, examination of the genetic and chromosomal health of the fetus, and fetoscopy, a technique that enables visual inspection of the fetus (*Fogarty International Center Proceedings*, 1970). Advances such as these are applicable only when a chromosomal or genetic defect has a structural or metabolic effect upon the developing fetus. Many genetic defects cannot yet be detected by these techniques. While amniocentesis appears to be used primarily for the detection of Down's Syndrome, fetoscopy is still in a largely experimental stage. These techniques are relatively expensive and not widely available.

The Role of the Genetic Counselor

Genetic counseling as a form of medical service differs from traditional medical practice. A complex relationship exists between the counselor and the counselee in genetic counseling. While some counselors perceive the relationship in the traditional doctor-patient manner, others perceive it as a counselor-client relationship (Sorenson, 1971b). Because the rights and obligations of both parties of this relationship have not been completely defined, the practice of counseling is probably based primarily on the beliefs, attitudes, and skills of the individual counselor.

Some counselors feel their primary task is simply to inform the counselees of the risks involved. They believe that to try to influence the decisions of the counselees is to go beyond the professional responsibility of the counselor. While a counselor may feel that the risk involved in a specific case is great or that a given disorder should not be transmitted to new carriers, he should refrain from telling the counselees what they should do. Often, in this situation, the counselor will define his relationship with the counselees as a counselor-client relationship (Reed, 1963). In so doing he is stressing the learning nature of the relationship. The client is there to be informed, not to be healed.

Characteristic of the traditional doctor-patient role is the healing relationship in which the counselee yields discretionary power to the doctor (Bloom, 1963). Some counselors describe their relationship with their counselees in this fashion (Lynch, 1969). The counselor may suggest how patients should use the information he gives them, and his concern for the patient goes beyond a mere statement of statistics of risk. For example, the counselor may discuss in detail the psychological consequences of an abnormal child, or the economic burden of genetic abnormality. He may also point out to the parents the impact on the family of a defective child. Finally, he may discuss with them the problem of dealing with an afflicted dependent when they advance in age and retire (Fletcher, 1971). All of these are important factors in conveying the meaning of genetic risks to parents and undoubtedly have an impact on the parents' decisions about future reproductive behavior.

Because his role falls within the purview of medical service, the counselor is probably approached by most people in the traditional patient manner. This orientation requires the counselor to assume much more discretionary power than most want. The meaning of a risk is complex, as is the meaning of disease or abnormality. While

179

many counselors and counselees may agree that it is good to avoid having mongoloid children, there is less agreement as one moves from physiological abnormalities to such problems as an XYY chromosome anomaly and the hypothesized link with social pathological behavior (Public Health Service, 1970).

The information the counselor gives, as well as the way in which he gives it, is crucial in defining the situation for parents. For example, if the counselor indicates that in the United States about 1 in 50 births will result in a severely abnormal child (Carter, 1969), he can label the risk high or low. If he says that the odds for a *normal* birth are 3 in 4, or conversely the odds for an *abnormal* birth are 1 in 4, the facts are the same but the impact of the information on the parents is likely to be reversed. Because counselors are generally aware of the problems of giving genetic counsel, many are concerned about exerting too much influence on the parents.

The Delivery of Genetic Counseling

The practice of genetic counseling appears to be moving from academic departments to departments in medical schools and in public health centers (Sorenson, 1971b). Today, the typical counselor is more often an M.D. or an M.D.-Ph.D. than the traditional Ph.D. geneticist of the past. These shifts have numerous implications for both the delivery and the scope of genetic counseling.

Whether counseling facilities are located in medical schools or public health clinics makes a difference in the distribution and utilization of counseling. Most doctors in medical schools occupy the dual role of teacher and researcher. Research interests are normally pursued by intensive study of a particular facet of disease. For example, if a doctor-researcher interested in genetics focuses his attention on cystic fibrosis, he will become professionally identified through publications and more informal channels as a specialist in cystic fibrosis. As a consequence, both the dictates of his research and his professional identification will restrict the types of patients he counsels. In addition, if the medical school in which the researcher works does not have a large department or program in human genetics, as is often the case, it will become identified as specializing in counseling on limited genetic topics. Patients with other genetic problems will not be referred to it and must go elsewhere for counseling.

Genetic counseling is beginning to be practiced in some public health centers (*Symposium on Human Genetics in Public Health,* 1964). With this development, the proportion of lower socio-economic groups receiving counseling will most likely increase. Public health nurses, whose professional duties are primarily concerned with lower-income groups, will become a major force in the dissemination and application of genetic counseling. The public health nurse can provide the family with attention at home, a service seldom available in genetic counseling today (*Symposium on Human Genetics in Public Health,* 1964). Genetic problems can, and often do, have a profound impact upon the social-psychological stability of the family (Zuk, 1959). Family follow-ups and counseling should provide for more effective genetic guidance.

Reproductive Decisions After Counseling

There is limited information on reproductive decisions made by patients after counseling. Certainly many factors are important in determining these decisions: (1) the size of the risk, (2) the severity of the potential abnormality, (3) the social and private attitudes of the parents toward abnormality, (4) the economic capacity of the family to endure the burden of a genetic disease, (5) the genetic health of existing childern, and (6) the type of counseling parents receive. In a recent study, Carter (1966) presented a follow-up of the reproductive behavior of 169 couples referred to a genetic counseling clinic in England. His study revealed that the magnitude of the risk had a large impact on the future reproductive behavior of the couples. Fully two-thirds of the high-risk parents, as against one-fourth of the low-risk group, did not have additional children. (Of course, factors other than the magnitude of the risk affect these reproductive decisions.)

As we have seen, couples who are faced with the prospect of a high risk of a serious defect may decide to take the risk and have a baby. If they decide not to take the risk, however, several options are available to them: (1) they may give up plans for having additional children; (2) depending on the nature of the genetic disorder, they may decide on artificial insemination from a donor; or (3) they may decide to adopt a child. As with the decision whether or not to take a risk, decisions among alternative ways to have a child are influenced by many factors. Unfortunately, there are no published data either on these decisions or on the effects upon the decisions of social class, religious beliefs, legal constraints, and financial conditions. Carter (1966) does indicate that within his sample of high-risk couples who decided not to have any additional children adoption was low. In several cases the wife was sterilized, and in many cases the husband.

With few exceptions, much of the preceding information is based on the personal experiences of genetic counselors. While this information is important, more systematically gathered knowledge is needed. The increasingly important role that genetics will play in medicine in the future requires that we understand how people react to and use genetic knowledge, whether they are professional medical personnel or the public. In addition, the important social, ethical, and legal problems beginning to arise from the development of new medical ideas and technology require solid knowledge of the extent and use of genetic counseling today. Knowledge is being applied and precedents are being set that will shape the use of future discoveries.

Moral and Legal Decisions in Reproductive and Genetic Engineering

WERNER G. HEIM

Man has long known that he can modify the genetic bases of organisms by controlling their breeding patterns. The application of such knowledge to the development of specific breeds of dogs or of wheat strains particularly suitable to cultivation is quite old. Mendelian and post-Mendelian genetics increased this ability by providing a rational model for the observed effects and thereby increased the ability to predict. This ability, when combined with advanced knowledge in the physical sciences, led to an understanding of many of the major mechanisms for the genetic transmission of information. Finally, this newest knowledge is rapidly providing the means for the manipulation of that hereditary information itself.

Imminence of the Developments

The ability to change the hereditary information content of a human cell is not a matter for the future: it is presently available. In 1971 Merrill, Geier, and Petricciani showed that a gene—a unit of hereditary information—from a bacterium could not only be introduced into human cells but could be caused to

THE AMERICAN BIOLOGY TEACHER, 1972, Vol. 34, pp. 315-318.

function in them. Nor need we any longer simply search for a gene that we may wish to manipulate: in 1972 a human gene was synthesized by Kacian and his associates. These techniques are still laboratory exercises; but the time-span from first demonstration to practical application is not likely to be longer than for similarly important developments in the physical sciences. Transistors, for instance, were in general use within less than a decade of their initial development.

An additional line of research is likely to accelerate the application of genetic manipulation, or genetic engineering, to man. This is what may be termed reproductive engineering. Reproductive engineering is any deliberate manipulation of the procreative part of the life cycle. Much of this is already in daily use in such forms as conception control and artificial insemination.

Genetic and reproductive engineering tend to be synergistic. Consider, for example, the production of genetically mosaic mice; that is, mice whose heredity is literally a patchwork. This condition is achieved by fusion of very early embryos of different strains. This requires at least four interacting techniques: (i) the exact timing of pregnancies, (ii) the manipulation of the eight-cell embryos in vitro, (iii) the production of exactly timed pseudopregnancies in the recipients, and (iv) the safe implantation of the mosaic, tetraparental embryos into the pseudopregnant females. The technique of transferring very early embryos from the uterus of one female to the uterus of another—this is called inovulation—itself has implications, to be discussed later. Another potential synergism between the two kinds of engineering may soon appear in the form of the insertion of genes into sperm prior to the use of that sperm in artificial insemination.

A warning should be added here, however. In all of these processes highly specific techniques must be used. This means that one kind of manipulation may be possible at a particular time while another, and apparently closely related, procedure remains out of reach. For example, it is now possible to insert a gene into a human cell; but to remove a specific gene is not yet possible, and indications are that it may remain impossible for a long time. It follows that only a person well acquainted with both the scientific literature and the conceptual basis of these activities is in any position to estimate the time scale for a future development.

Uniqueness of the Developments

These new scientific developments, which have been and are being incorporated into the technology of our society, pose serious moral and legal questions. That, of course, is not in itself new; scientific developments, when more or less broadly applied, have always raised such questions. (Recent examples include the fluoridation controversy and the legal aspects of artificial insemination.) What is new is the nature of the questions raised by the development of genetic and reproductive engineering: the manipulations of man, in these fields, will be vastly more fundamental than any previous ones. In some instances the manipulations will be irreversible, not only with respect to the individual but also with respect to all his descendants and to the population. In such circumstances wrong decisions can lead not only to the death of the individual but also to extinction of the species. Other incorrect decisions could seriously degrade the quality of life through control of thought, of liberty, and of motivation.

Roles in Controlling the Applications

Unfortunately the track record of those who, in my opinion, should be developing the moral positions, public policies, and legal constraints concerning these developments is not good. With few exceptions the moralists, theologians, sociologists, political scientists, legislators, legal scholars, and judges, whose job it is to integrate new developments into a sustained, viable, and sound fabric of society, have not only failed to be prepared for the introduction of new, science-derived technologies into the culture but have even reacted, at best, rather sluggishly. Their reaction time has sometimes been slower than that of the general public, as is shown by the widespread acceptance of certain developments long before they are placed in a definite legal and moral framework. An example is the debate about "when and whether to pull the plug" of the machine keeping a terminal patient more or less alive. This debate

185

commenced seriously only well *after* the machines had come into common use and is still often being conducted in terms hardly useful to the person who must make the decision: the physician on the ward. Another example of the failure to develop an adequate set of moral stances has to do with genetic counseling and the individual and societal risks arising from the reproduction of persons who are sustained by medical skills in the face of hereditary defects. Should an 11th commandment have been promulgated: "Breed not, ye who carry defects"?

One popular reaction to this lack of guidance has been a feeling that research in these fields should be stopped because "it's too dangerous." Of course it is not the research itself that is dangerous but the application of the results within a society that has developed neither the broad principles nor the specific directives to make these applications wisely.

By default, the natural scientist himself has had to fill the vacuum with his own inexpert opinion so frequently that the impression is now abroad in some circles that this, too, is a part of his job. The job of the natural scientist is to make the discoveries; that of the technologist is to develop applications; and that of the social scientist and humanist is to suggest whether, how, and under what conditions the work of the other two ought to be applied.

As a well-informed citizen the natural scientist should have, of course, the same opportunity and at least the same responsibility to contribute to the decision-making process as any other well-informed citizen. But he is not an expert in, and should not be expected to act as if he were an expert in, the delicate processes of weighing the data or of drawing **conclusions and making recommendations. He does, however, have responsibilities in this constellation of events beyond those of the ordinary person. First,** he should give plenty of advance notice to the humanists and social scientists of forthcoming developments likely to require their attention. In doing so he must put all his expertise to work to differentiate between science and science fiction, lest the world react like those to whom the shepherd cried "wolf" too often. Second, he must be prepared to give the relevant details of new developments to his colleagues in the humanities and the social sciences. He must do this with due regard for (i) relevancy, lest he either swamp or impoverish the communication channels; (ii) scrupulous accuracy, lest he mislead his listeners; and (iii) intelligibility of language, lest he be misunderstood. Finally, he ought to monitor the tentative and the definitive pronouncements of the humanists and social scientists so as to detect early any misinterpretation or lack of facts that may have distorted their work.

The need for careful formulation of moral and legal positions on new developments *before* their widespread use is now more critical than ever before: the changes are more fundamental in nature, are less likely to be reversible in the individual or in his descendants, and, most importantly, are changes in human nature itself. As Leon Kass (1971) has pointed out, ". . . both those who welcome and those who fear the advent of 'human engineering' ground their hopes and fears in the same prospect: *that man can for the first time recreate himself.* Engineering the engineer seems to differ in kind from engineering his engine." (Kass's emphasis.)

Aspects of Human Engineering

The "human engineering" of which Kass speaks— it is inclusive of what is here called genetic and reproductive engineering—differs from previous kinds of changes in several respects. Consider its influence on human conduct. The traditional methods of modification have had three important characteristics: (i) they used symbols, especially as embodied in speech and art, as their primary vehicle; (ii) they allowed considerable choice to the individual as to the acceptance of at least parts of the modifications offered; and (iii) their effects could, to a great degree, be reversed in both the individual and his progeny. In contrast, the changes brought about by human engineering are largely nonsymbolic, because they modify the conduct-controlling mechanism directly. For example, the human engineer would *not* seek to educate or train a victim of Down's syndrome (the so-called mongolian idiot) to the point of self-sufficiency but rather would eliminate the chromosomal defect or, alternatively, would block fertilization or development of eggs carrying the defect. This is, incidentally, in sharp contrast with one of the most advanced of the old-style techniques, Skinnerian operant conditioning.

The second contrast between the old and the new techniques lies in the fact that the individual frequently will have no power to adopt or reject a particular modification. The modification will have occurred before he became an individual—either in the gametes or the early embryo that produced him or in an ancestral generation. To the extent that the change is in the hereditary material itself, it may be an irreversible change for either of two reasons: the technique for reversing it may not be available— at present a gene may be added but not subtracted— or else the desire to make the reversal at some future date may have been blocked genetically at the same time that the other changes were made.

A further consequence of the difference between the old and the new techniques is that the new techniques have the effect of removing literature, mythology, religious liturgy, etc., one more step from the real seat of power. Once, long ago, a speech—a curse—was expected to kill an enemy. Later a speech was expected to inspire soldiers to kill the enemy. In the future a speech might cause technicians to change the genes of some persons so that, under certain conditions or at a certain age, these persons simply die.

"Now" and "Soon" Examples

What, then, are some of the more likely of these human-engineering techniques? I wish here to deal only with those that are either available right now or have a great likelihood of becoming available in the next few years. Due to the highly technical nature of the work, long-range prophecy is likely to be unprofitable.

One technique that fits the human-engineering category, although it is neither reproductive nor genetic, is that of the direct control of man's neural system by electrical or drug stimulation of specific parts of the brain, as detailed so well by Delgado (1969). One need only read his book to discover the tremendous power available here and the need to develop a proper framework for that power.

Other nonreproductive and nongenetic methods are those devised for the prolongation of life by what, at any point in time, would be considered "extraordinary" means: heart transplantation, heart

188

prosthesis, and the like.

In the realm of genetic engineering itself, one must at least mention genetic counseling—already widespread and rapidly spreading—because serious moral problems crop up almost daily in this work. If a prospective child has a risk of 5% that he will be seriously defective, should the parents reproduce? Or, if conception has occurred, should an abortion be done? What if the risk is 10% or 30% or 60%? What if the risk is essentially zero that the child will be affected but quite substantial that the child will transmit the deleterious gene?

Let us turn now to questions of genetic engineering in the strictest sense. Here we are dealing with techniques that change "human nature" and are transmissible directly to the offspring. Under what conditions should such techniques be used? The technique for inserting a gene or a group of genes into sperm is likely to be practical within a very few years. Suppose two persons are defective in the gene for the production of the enzyme parahydroxylase: their child may suffer from the disease phenylketonuria (PKU). It should be possible soon to introduce the normal, active gene into the sperm of the father and thereby assure that the cells of the offspring have, each, at least one properly functioning gene for this enzyme. Should this procedure be done? Consider not only that normal offspring will be produced but that both of the defective genes will be available to, although perhaps not active in, further progeny.

In the eugenic uses of genetic engineering that I have mentioned here, most of the arguments seem to be on one side: because it is highly likely that both the individual and society will benefit from the engineered changes, and because either short-term or long-term deleterious effects are unlikely, the judgment probably will be that the procedure should be carried out. But let's take a less simple case. There is good evidence for a hereditary component in the origin of schizophrenia and schizoid conditions (Heston, 1970). Very likely we are dealing with a multigenic inheritance pattern here. Therefore there may be a defective gene that facilitates the appearance of a mild schizoid condition. It is not unreasonable to suppose that it should become possible to insert a gene into the fertilized egg—a gene that would "cor-

189

rect" the defective gene. At first sight this would seem to be another clear-cut case for the use of the technique. But there are reasons to suppose that mild schizoid tendencies are useful in a wide range of endeavors, from painting and composing to scientific research. It is obvious that, without *much* further information, both the long-term and the short-term benefits cannot be weighed properly against the possible harm. And remember that once such a corrective gene has been introduced, it may not be removable! In such an equivocal situation what ought to be the rules for the application of the technique? I submit that, in such a situation, the role of the biologist is to provide the raw information and, perhaps, an indication of the probabilities of the various consequences. Then the humanists and social scientists should formulate the rules, subject to revision as more information or better techniques become available.

The other type of human engineering—reproductive engineering—also will require much thought and study. Although less likely to have permanent or irreversible effects, its points of action are even more closely bound up with our general ideas of morality and proper conduct than are those of genetic engineering. This is evident from the arguments raging around the two early forms of reproductive engineering presently in use: conception control and artificial insemination. Neither legal nor moral opinion has been stabilized or universalized in respect to these. What are and what should be the legal and moral positions vis-à-vis the following procedures, all but one of which have already been carried out in nonhuman organisms?

1. Artificial insemination with sperm from donors selected for particular qualities.

2. Artificial inovulation in which an egg produced by the wife and fertilized by the husband is transferred to a foster uterus.

3. Artificial inovulation by transfer of an egg produced by a donor, and fertilized either by the husband or a sperm donor, into the uterus of the wife.

4. Insertion of the nuclei of ordinary body-cells into eggs whose own nuclei have been removed, producing thereby any desired number of individuals with exactly the same genetic constitution as that of the donor of the nuclei—a procedure called cloning.

5. Cloning by causing ordinary body-cells to act as if they were fertilized eggs, thereby (again) producing any desired number of persons having the same hereditary makeup as the donor of the cells.

So little thinking has been done about the last two of these possibilities (outside the realm of science fiction, anyway) that we can hardly even list the individual or social advantages and disadvantages that might accrue from having a large number of genetically identical individuals. We are even less in a position to give these possibilities the calm and deliberate weighing that ought to lead to acceptable patterns of application.

Prospects and Challenges

The following general conclusions can be drawn:

1. Specific modification of individuals, by action on the genes or on the reproductive patterns and for individual or social purposes, is now possible and is increasingly and rapidly becoming feasible.

2. These techniques differ fundamentally from older approaches based on restricted breeding patterns—classical eugenics and Hitlerian eugenics, for example—in two critical respects: they are fast and they work. Even the insertion into man's heredity of what might be termed socially specific genes is a rather close—an uncomfortably close—probability.

3. We need to develop ways of adjusting these possibilities, of restricting their use within morally and socially acceptable patterns.

4. To do this will require a rejuvenation of the humanities and the social sciences and a reshaping of the relationship between them and the natural sciences.

Athelstan Spilhaus (1972) has said: "Just as technological invention cannot remove the need for social invention, neither should our slowness in changing outmoded social practices, institutions, and traditions be allowed to slow technological realization of potential benefits to all." Unless we rearrange our houses so that we can get to work immediately and unless we *do* get to work immediately, the genie (or **gene) will be out of the bottle before we know the magic formulas that control it. And to permit this kind of genie to be out of control or to be misused is (to use the words in an unusual but meaningful** combination) to *commit extinction*—first of freedom, and then of the species.

As teachers we have additional duties: to inform our students of these developments and to help them make the questions raised by these developments a prime order of business in their lives. We may confidently expect their lives to encompass the period during which most or all of the developments outlined here, as well as others, will be brought into practice. Those who are presently our students will be the ministers, the judges, the political scientists, the philosophers, the legislators, the teachers and, yes, the biologists, in charge. We and they had better start to think fast about how to handle that charge.

REFERENCES

DELGADO, J. M. R. 1969. *Physical control of the mind: toward a psychocivilized society.* Harper & Row, New York.

HESTON, L. L. 1970. The genetics of schizophrenic and schizoid disease. *Science* 167: 249-256.

KACIAN, D. L., *et al.* 1972. In vitro synthesis of DNA components of human genes for globins. *Nature New Biology* 235: 167-169.

KASS, L. R. 1971. The new biology: what price relieving man's estate? *Science* 174: 779-787.

MERRILL, C. R., M. R. GEIER, and J. C. PETRICCIANI. 1971. Bacterial virus gene expression in human cells. *Nature* 233: 398-400.

MINTZ, B. 1971. Genetic mosaicism in vivo: development and disease in allophenic mice. *Federation Proceedings, Federation of American Societies for Experimental Biology* 30: 935-943.

SPILHAUS, A. 1972. Ecolibrium. *Science* 175: 711-715.

GENETIC ENGINEERING

John V. Tunney
Meldon E. Levine

The cry has been raised by many that the impact of science has been too fruitful. It has been raised by some with regard to the nuclear sciences. It might well be reiterated in the near future with regard to the biomedical sciences.

The biomedical sciences have provided man with the increasing ability to modify and to alter human genes. These developments touch upon the most fundamental issues of human life. They portend the ability to reshape man.

Accepted attitudes about the inviolable nature of man's genetic endowment now stand challenged by science. The political impact of this challenge might be just as powerful as the scientific impact. Unless the potential provided by the biomedical sciences social response might be fashioned out of fear. When complex problems appear too terrifying or mysterious, some people might seek simple solutions that prove inadequate or improper.

If people become sufficiently frightened—if they feel the need to be rescued from a menace they do not understand—they are more likely to delegate freedoms and less likely to respond with reason. If the polity responds to the scientific community through fear and mistrust, we could witness the erosion of our most precious freedoms. Political alteration—like genetic alteration—might be irreversible. If our personal liberty is ever lost, it might never be recovered.

Consequently, the most important and enduring of our freedoms are linked with the manner in which the biomedical sciences are understood and applied. The issues raised by the biomedical sciences must be exposed to public scrutiny. They must be discussed candidly, openly, and at once.

In approaching these issues, we the authors must of necessity wear several "hats." The first "hat," if you will, is a multifaceted one, fashioned around our own personal backgrounds—our social, philosophical, and ethical beliefs. The second "hat" is a legal one, obtained after studying the law and participating in it over a period of years. The third is a legislative one, obtained from the unique perspective to which we are exposed in helping to propose, evaluate, and create the laws of this land.

It is important to recognize that all political figures wear "hats" of this nature. They are all different, depending upon the individual background and experience of the person, but they influence him as he evaluates and determines policy, especially in an area as sensitive and as potentially explosive as genetic engineering.

Before we discuss the ethical, legal,

SATURDAY REVIEW, August 5, 1972, pp. 23-28.

193

or legislative view, however, it is important to set forth the most salient aspects of genetic engineering and to indicate our assessment of the state of the art in each of these aspects.

ABORTION AND AMNIOCENTESIS. The technique of amniocentesis—prenatal sampling of the amniotic fluid surrounding the fetus—is frequently used to provide advice on therapeutic abortions. The procedure is relatively safe, but we do not yet have the ability, with amniocentesis, to detect all genetic defects. Within five years most monogenic defects that we understand will be detectable thereby, but even then questions will remain unanswered as to whether amniocentesis affects the eventual intelligence of the child.

MASS GENETIC SCREENING. In this, too, we appear to be on the threshold. The technique is available for many diseases, although not for some others. However, it has already become evident that many people will oppose mass genetic screening—whether of children or adults—for a variety of personal reasons. Some feel it is an invasion of personal rights; others do not want children genetically defective to find out that they are so afflicted. Again, however, the technology is increasingly available.

MONOGENIC GENE THERAPY. Modification of certain cells in terms of their genes, or monogenic gene therapy, we have been advised, has not yet been performed successfully. It should, however, be a possibility for certain diseases within five years. As we gain more knowledge about monogenic defects, the possibility of monogenic gene therapy will become more of a reality in broader areas.

IN VITRO FERTILIZATION. Both in vitro fertilization and reimplantation in the uterus have been performed successfully in experimental animals. If the research in this area is not seriously inhibited by external controls, the technology for in vitro fertilization and for reimplantation in human beings should be available within five to ten years, or perhaps even earlier. Recently a Yugoslav scientist advised a conference in Tokyo that had developed an instrument for oöcyte (egg) transplant into the uterus.

CLONING. Cloning of frogs, where a replica of an individual is developed from one of its somatic cells, has already been successful. The technology for the cloning of mammals will be available within five years, and, unless research is stopped, the technology for the cloning of human beings might be available within anything from ten to twenty-five years.

POLYGENIC GENE THERAPY. We are very far away from achieving polygenic gene therapy—perhaps 50 to 100 years. Our understanding of polygenic gene defects still is extremely primitive. For a variety of reasons it is considerably more complicated to isolate and trace a polygenic trait than to isolate and trace a monogenic trait.

One of the most powerful arguments presented in favor of employing one or more of the technologies of genetic engineering in the direction of genetic intervention is that man's genetic load is increasing. In other words, the total number of genetic defects carried by man has been increasing. This has been occurring as a result of the increased mutation rate that accompanies population growth and the decreased natural selection rate occasioned by modern medicine and technology. About one child in twenty, for example, is now said to be born with a discernible genetic defect.

The question then arises: Should the human species attempt to employ these new technologies to deal with this increased genetic load? Paul Ramsey of Yale has stated that "it is no answer to say that changes are already taking place in humankind or that men are constantly modifying themselves by changes now consciously or unconsciously introduced. . . ." He argues, as a Protestant theologian, that scientific intervention in this area is a questionable human aspiration, as he puts it, "to Godhood." Regardless of the merits of Ramsey's position, the very vigor with which he defends it suggests the extent to which ethical issues are at stake.

The ethical questions raised by the possibilities implicit in genetic engineering are no less fundamental than the issues of free choice, the quality of life, the community of man, and the future of man himself. Thus, it becomes evident that one's own sense of ethics, one's personal view of right and wrong, one's own standard of conduct or moral code, are essential components of decision making in this extremely sensitive area.

Many political scientists like to believe that political decision making can be objectified, that a process can be delineated by which political decisions are made. Through such a process, it is assumed, decisions and actions can be predicted. The wisdom of these political scientists is questionable for a variety of reasons. One of the most important is the significant subjective component of political decision making—the large realm left to one's own values and ethics.

This realm affects all aspects of lawmaking. It is especially important in any political or even legal approach to genetic engineering. One's own values or ethics must inevitably be brought to bear upon a variety of important questions in this area, questions that can be evaluated only by subjective criteria. In an effort to rationalize some of the issues involved, we will attempt to draw some distinctions and to articulate some criteria for analysis. Some of these criteria have been suggested by others; some are our own. We suggest them not as a definitive list but as a reminder that it will be very important to apply criteria such as these to any legislative or legal analysis of the implications of genetic engineering.

Let us posit a list of ten general considerations suggesting possible ethical distinctions:

First, if we are to engage in any eugenics, negative or positive, we must confront three vital questions that pervade this entire subject: What traits are to be considered desirable? Who is to make that determination? When in the course of human development will the choice be made? These questions cannot be underestimated in their importance to the future of man, particularly when we are considering biological alternatives that might not be reversible.

Second, we must ask whether the genetic engineering or "improvement" of man would affect the degree of diversity among men. Does it presume a concept of "optimum" man? Is diversity important as a goal in itself? Does —or should—man seek an "optimum," or does he seek a "unique"? What would the quest for an "optimum" do for our sense of tolerance of the imperfect? Is "tolerance" a value to be cherished?

Third, we should consider whether it might be appropriate to delineate different biological times or moments— at least in humans—during which experimentation might occur. Do different ethical considerations apply if we attempt to distinguish between experimentation on an unfertilized sperm or egg, a fertilized sperm or egg, a fetus, an infant, a child, or an adult? Might the factors to be balanced in making a decision as to whether experimentation is proper vary at different stages of human development?

Fourth, is there a workable difference between, on the one hand, genetic "therapy" to correct genetic factors known to cause somatic disease and, on the other hand, genetic "engineering," defined as techniques to alter man in terms of some parameters other than somatic disease? Might it be appropriate to attempt such a distinction in definitions in this emotionally charged area? Might the term "genetic therapy" evoke less emotionally charged reactions than the term "genetic engineering"? Might it, in fact, be preferable to respond more receptively to those areas of genetic work that are primarily "therapeutic"? Or is such a distinction unworkable?

Fifth, it would seem to be advisable to ask whether a particular technique or technology is devised for the therapeutic treatment of an individual or whether it is designed to have a broader societal impact. This potential distinction has a variety of ramifications. For example, it should be asked whether techniques developed for the therapy of an individual patient automatically diffuse into the general public for purposes other than this therapy. Are physicians operationally capable of restricting the use to one group, or does societal pressure make them semiautomatic dispensers of seemingly desirable technologies?

Sixth, we might ask whether any eugenics program—whether positive or negative, voluntary or compulsory—does not imply a certain attitude toward "normalcy," toward a proper norm for human activity and behavior, and toward expectations with regard to the behavior of future generations of human beings. Implicit in this question are distinctions with regard to positive versus negative eugenics programs and also with regard to compulsory versus voluntary eugenics programs.

Seventh, how are words such as "normal," "abnormal," "health," "disease," and "improvement" defined? Are they words that can be operationally used to determine what should be done in the area of genetic engineering?

Eighth, we must ask if the quest for genetic improvement would be continuous. Would it invariably make all children "superior" to their parents? What would be the social consequences of this? Would it institutionalize generation gaps and isolate communities by generations?

Ninth, we should consider whether the institutionalization of a quest for genetic improvement of man is likely to lead to his perception of himself as lacking any worth in the state in which he is. What does this do to the concept of the dignity of the human being in his or her own right, regardless of some "index of performance"?

Tenth, if we have a well-developed ability to perform genetic therapy as an assault upon certain diseases but such therapy is not available for all who have the affliction or who desire the "cure," the question will immediately arise as to how to determine which patients will receive it. Are some classes or groups of people more desirable patients or more worthy of treatment? How will selection be made? By what criteria will those decisions be reached?

Questions such as these ten can be answered only by appealing to ethical, or so-called moral, arguments. When we enter this realm, it is important to remember that no one has a greater claim to wisdom than anyone else. All men have a stake in this area, and all men have a right to be heard.

We would like to offer three additional thoughts that might affect all of the ethical judgments involved. Two are caveats, and one might be a preliminary guide for analysis.

The two caveats are reminders of the imprecision of measurement and the difficulty of meaningful analysis in this area. As for the imprecision of measurement (caveat number one), Ramsey states that "many or most of the proposals we are examining are exercises in 'what to do when you don't know the names of the variables.'" While that might be somewhat harsh, he is accurate in his suggestion that prediction of behavior or even of most genetic disease will be very difficult, owing not only to polygenic factors but also to such other imponderables as pinpointing a recessive trait in its heterozygous state and predicting the influence of environmental factors.

The second caveat is the difficulty of meaningful analysis. Some values we will be asked to compare will be like comparing apples and oranges. How

196

can one, for example, compare the possible deep satisfaction experienced by an infertile woman carrying and bearing a child that was fertilized in vitro and reimplanted in her uterus with the 1 or 2 or 5 per cent chance that the child will be deformed? In measuring eugenic traits to be cherished, how can one compare intelligence (even assuming it can be defined) with love?

In making genetic choices—and in selecting those who will make them— one should not forget such caveats.

The last of these attempts at ethical classification is an effort to ask the question of just where in the broad field of genetic engineering the ethical issues will arise. At what level in the process? Professor Abram Chayes of Harvard Law School has suggested that at least three levels can be discerned at which the questions posed above might arise:

First, the general level of research. Should research be pursued that might lead to technologies that will give science the genetic capability to engineer human beings?

Second, the level of treating human disease. Questions will, of course, be raised as to what exactly is a disease— and how it is defined. (Should socially undesirable or disruptive behavior be treated as an illness? Are some forms of mental illness proper candidates for genetic therapy?) Even assuming that those questions can be answered, ethical considerations will arise with regard to whether the disease should be treated with a newly available genetic technique.

Third, the broad level of attempting to affect society—the level of what some will consider an improvement of the human species.

These levels overlap to some degree with the questions we have already raised, and it appears clear that the ethical pressures that will be directed against the continuation of the activity will become increasingly strong as we move from the first to the second and then to the third level. While these considerations by no means exhaust the ethical realm, they do suggest the enormity of the problems with which we are attempting to deal. Perhaps the attempt—however primitive—at ethical classification might also offer the lawyer some general guidance.

It might be asked, for example, whether there are legal as well as ethical distinctions between negative and positive eugenics. Are there legal differences between an attempt, on the one hand, to treat an individual for disease by either monogenic or polygenic gene therapy and an attempt, on the other hand, to control behavior or otherwise alter society's norms? Does it matter—legally—at what point in the state of human development the therapy or the engineering is conducted: whether in the stage of birth control, in the realm of abortion, in treating a minor or an adult? Does it matter—legally—how the therapy or the engineering is conducted, whether it is voluntary or compulsory, whether for punitive or eugenic reasons, or whether the physician has freely and openly obtained the consent of the patient?

Clearly, these distinctions ought to be important—legally as well as ethically. They touch upon fundamental and traditional legal principles, principles that have been applied in Anglo-American jurisprudence for a number of years. They offer the lawyer a variety of factors that will help in his analysis.

Law, at least in the United States, can be said to operate on three broad tiers or levels. First, we have constitutional law, or the legal framework set forth by the Constitution of the United States and by the courts in interpreting the Constitution. Second, we have statutory law, or law that is enacted by statute—either of a state or of the federal government. No statute can contravene a constitutional requirement. But, in the absence of a constitutional prohibition or pre-emption, federal and state legislatures can enact statutory standards to respond to a variety of needs, such as those that arise in the areas of health and welfare. Third, in the absence of a controlling constitutional or statutory provision, the courts rely upon the body of law known as the common law—those legal principles that have emerged from judicial decisions. The issues raised by the technology of genetic engineering affect constitutional, statutory, and common-law principles. We shall briefly consider each of these legal tiers.

At least three constitutional factors clearly emerge when one considers the general subject of genetic engineering. The first is the right to privacy. The Fourth Amendment to the Constitution declares that "the right of the people to be secure in their persons, houses, papers, and effects, against unreasonable searches and seizures, shall not be violated. . . ." This language has been interpreted to guarantee to the individual a constitutional right of privacy. Genetic engineering raises questions with regard to the extent and inviolability of that right. The second factor involves the rights protected in the Fifth and Fourteenth amendments, which guarantee that no person shall be deprived of life, liberty, or property without due process of law. Third, and perhaps the most important factor that the Constitution brings to bear upon genetic engineering, is the *approach* of constitutional law—the method of analysis that courts have developed for dealing with constitutional issues. Apart from the technicalities inherent in whether state action is or is not involved—a threshold question in any constitutional analysis—constitutional law requires the government to show a more compelling governmental need when the abridgment of fundamental freedoms is involved.

Let us take two examples. Contrast, for instance, a government-sponsored compulsory program of negative eugenics, designed to eliminate a certain genetic disease, with a government-sponsored compulsory program of positive eugenics, designed to control behavior. As both programs are compulsory, both could infringe the fundamental freedom of procreation and possibly of marriage. However, compelling state interest could be advanced as a more legitimate argument in eliminating a disease rather than in controlling or altering behavior. The eradication of disease has long been accepted as a vital social objective. We do not offer this dichotomy in an effort to support the negative eugenics program. In fact, we would probably oppose it. But we do think that a constitutional analysis of the two approaches would bring different factors into being and might yield different results in the two cases.

To move from constitutional law to statutory law, it should be noted at the outset that a variety of statutes in numerous American jurisdictions have attempted to impose eugenics controls. Professor William Vukowich of Georgetown has written: "In the early 1900s, many states enacted laws that prohibited marriage by criminals, alcoholics, imbeciles, feebleminded persons, and the insane. Today most states prohibit marriage by persons with venereal disease but only a few states have laws which are similar to those of the early 1900s. Washington and North Dakota, however, still prohibit marriage by women under forty-five and men of any age, unless they marry women over forty-five, if they are an imbecile, insane, a habitual criminal, a common drunkard, feebleminded or [a] person who has . . . been afflicted with hereditary insanity."

A number of the more recent developments in the field of genetic engineering, however, go entirely unregulated. Sperm banks—which may be used as a reserve of sperm for artificial insemination by third-party donors, for example—are an excellent example of institutions for which pertinent statutes do not exist. Their administration is entirely up to the persons operating them.

That is an instance in which our third legal tier, the tier of the common law, must be our guide. In the absence of constitutional or statutory guidance, we must turn to the common law for our standards. Here again the law is neither silent nor comprehensive. It falls somewhere in between. Assume this set of possibilities: Amniocentesis is an everyday practice, held by most doctors to be free of harmful effects such as infection. A woman who has not been offered amniocentesis gives birth to a Mongoloid child. Is her obstetrician liable for malpractice?

Common-law tort principles of malpractice would probably hold that the doctor would, in fact, be liable. This is so because the common law in determining negligence tends to follow whatever is the accepted medical practice for a particular community. But is this a viable solution? Would it be appropriate to require amniocentesis even if the mother—or the doctor—has strong religious convictions that preclude consideration of an abortion under any circumstances? What about offering amniocentesis under those circumstances? And what about the legal rights of an egg that has been fertilized and grown in a test tube? Does the *father* have any rights? What rights does the *mother* have? Or the doctor? Do common-law tort or property rights apply to this question?

However one evaluates these issues, they must be faced. If one does not wish to face them with the exclusive guidance of the common law, the result will be the consideration of new legislation. To the extent that current law is inadequate, legislation must be developed.

In considering the possibility that legislation must be developed, we both are painfully aware of the potential inadequacy of the legal and the legislative processes in responding to issues presented by science. In the area of genetic engineering, science may be outpacing the legal and legislative processes. It may be presenting challenges to which our lawyers and legislators are ill-equipped to respond. Our legislative system may be poorly equipped to respond to these problems because of at least two inherent difficulties: its speed and its scope.

Our legislative process generally works slowly. Sen. Walter Mondale, for example, first introduced legislation that called for a commission to study the effects of genetic engineering almost five years ago. That bill passed the Senate unanimously last year, but it has not yet been acted upon by the House of Representatives. Just initiating a study commission on so momentous a subject has already, then, taken longer than five years.

Not only is our political system slow. It is, obviously, only national in scope. Generally, that is not a significant problem to the people in Washington who are considering various legislative proposals. Most proposals are only national in scope—or less. Genetic engineering, however, is clearly a matter of international concern. It will require, if any controls or guidelines are to be effectively suggested, international agreements. This also will serve as a political or legislative constraint.

Recognizing these constraints, we still believe that certain constructive steps can be taken—steps that will begin to offer legislative rationalization to the field. If the legislative system begins to consider these problems now, it might be possible to respond politically and legislatively before it is too late.

Conversely, we fear the consequences that could be wrought if informed legislative consideration of the issues inherent in genetic engineering does not soon begin. If the legislation comes as a result of dramatic scientific breakthroughs that scare the public, the outcome might be hasty and unwise political decisions predicated upon inadequate information and upon fear. If debates and discussions begin now, however, the ultimate legislation might emanate from deliberate and reasoned political, social, and scientific analysis. We do not believe that, at this point, it would be appropriate to suggest answers to the momentous issues. raised by genetic engineering. But we do believe that we know enough to undertake certain legislative initiatives. Let us suggest three.

First, Congress should enact the Mondale bill [S. J. Res. 75], which provides for a study and evaluation of the ethical, social, and legal implications of advances in biomedical research and technology. The proposed study commission might serve as a preliminary vehicle for educating the public about the foreseeable social consequences of biological advances. Such a commission might best be an international one, but that is logically a second step.

Second, and perhaps equally important, is the initiation of technology assessment in all institutions that disburse funds, direct research, or provide grants that are related to biomedical concerns. It has long been obvious that technological developments have implications that affect society in a variety of ways and that their impact cannot be limited to an analysis of the technical aspects of the product or of the innovation. Similarly, the myriad implications of the developments of biomedical research reach out to all segments of society. Technological assessment should be a part of any analysis of any project that involves a potentially new biomedical development.

Third, it might be appropriate for Congress to earmark a small proportion of health research funds (say one-quarter or one-half of 1 per cent) for research into possible social consequences of biological technologies either presently available or foreseeable.

It is not only the legislature, however, that can initiate improvement in communication between the scientific community and the general public as well as expansion of public awareness of and concern with these issues. Four other suggestions might be worthy of consideration:

First, private foundations should be urged to initiate programs to bridge presently existing gaps between the sciences and the humanities, exposing people in each area to people in the other, and making the ideas of each readily available and understandable to the other.

Second, universities should consider establishing additional programs whereby students in the humanities would be exposed to the methodologies familiar to those in the scientific disciplines, and vice versa. The two general groups should feel a closer relation to and understanding of each other in universities as well as elsewhere. The effective separation of these two groups in universities—particularly at the graduate level, but even at the undergraduate level—establishes a line of demarcation between those in the sciences and those in the humanities, with very inadequate and narrow bridges to unite the two general areas.

Third, research proposals in the biological area should perhaps be assessed by institutional research review committees that include nonscientists. Some form of technological assessment, in other words, or consideration of the ethical, moral, and social implications of biological projects—by nonscientists—should be considered at the level of all research proposals in this general area.

Fourth, it might be appropriate for the medical profession itself to study the ways in which the technologies it uses for the benefit of individual patients may affect society as a whole if used for purposes other than the cure of individual patients. The "individual treatment versus social engineering" dichotomy should be considered clearly and carefully by the medical profession and should probably be emphasized more strongly than it currently is.

These efforts to bring society and the biomedical sciences closer together are, in our opinion, essential. Dr. Andre Hellegers, director of the Joseph and Rose Kennedy Institute for the Study of Human Reproduction and Bioethics of Georgetown University, has testified before the Senate that "nothing could be worse than that society should come to fear scientific progress. . . . I can foresee that the occasional, seemingly sensational, scientific episode will so frighten society as to undermine the very support [that] science needs in order to continue to make contributions to improve the lot of mankind. . . . It is high time that there be started an educative process that explains to the country the precise nature and limitations of the scientific process and the place it occupies in man's control of his environment. . . . No segment can stand apart in this interdependent society. If it attempts to do so, it is bound to cease being supported. The sooner the relationship of science to society is examined and explained for all to see, the better it will be both for science and for society."

This testimony touches upon two very important facts of American political life, neither of which should be forgotten. First is the theory of political accountability; if the public supports something financially, the public is entitled to know what it is that it is supporting. Second is the foundation of political democracy; thoughts, suggestions, proposals, and policies should be scrutinized in the market place of ideas. Political debate and public discussion are healthy and are conducive to the best analysis of any position. Particularly in an area as fraught with subjectivity as this one, it is vital that the issues raised be aired, discussed, and debated. We are dealing in an area in which there is no monopoly of expertise. Rather, it is a field in which men trained in a variety of different areas, or even in no special area, bring to bear their own unique perspective,

or, if you will, "expertise." We are dealing with a subject in which morality, or one's own subjective sense of ethics, is pervasive. We are, therefore, dealing with an area in which all persons have a right and a special claim to be heard.

There are certain suggestions that we would offer in any debate on this subject. We would suggest that among the values that man ought to protect most fully are the values of humility, of compassion, of diversity, and of skepticism. We would suggest that any scientific or technical initiatives of one generation that would foreclose or eliminate the options of future generations—any decision today that implies an ability to predict the human traits that will be most cherished tomorrow —smacks of arrogance and should be avoided. We would suggest that man should exercise the utmost caution in this sensitive field and that decisions that will be genetically irreversible might require a wisdom we do not possess. We would also suggest that there is no reason why the ethics or morality of any one of us is better than that of any other. In the realm of morality each of us has an equal claim to wisdom.

Therefore, the issues raised by the biomedical scientists must be debated, and the debate must begin now. If we postpone debate in this area, we might face irreversible trends not only in genetics but also in political freedoms.

All segments of society should be involved in the debate these new technologies demand. The techniques must be discussed and debated among lawyers, doctors, theologians, legislators, scientists, journalists, and all other segments of society. The issues raised require interdisciplinary attention. We cannot begin too soon to consider them.

INVIT: THE VIEW FROM THE GLASS OVIDUCT

The name has a vaguely Scandinavian sound with the accompanying promises of advanced sex. It is *Invit*, and it will be the result of the most advanced sex the planet has seen to date, though possibly not of the type you are thinking about.

Invit has been conceived not in a woman's body but in a laboratory apparatus. When he or she is born—and elements of the scientific community feel the birth will be within the next twelve months—Invit will be recorded as history's first known "ectoconceptus" (one conceived outside the womb), the progenitor of the test-tube baby.

Invit will be British. Disregard that misleading Scandinavian echo—the name is derived from the Latin *in vitro*, meaning a biological reaction taking place in an artificial apparatus, rather than within a living organism, *in vivo*.

The midwives are a group of intense and tight-lipped scientists and physicians led by Robert G. Edwards of Cambridge University's physiology laboratory. The actual birthplace, however, will be the Oldham General Hospital near Manchester, and the obstetrician-in-attendance is to be the hospital's Dr. Patrick Steptoe.

In the amphitheater (in thought at least) will be the world's most distinguished physiologists, including a vocal and highly critical group of Nobel laureates who believe that Edwards is committing an abominable act. James Watson of DNA fame, for example, has publicly demanded that Edwards abandon the Invit experiments. Max Perutz, also a laureate and a senior scientist for the British government's Cambridge-based Medical Research Council, has called Edwards's work a "stunt" and believes that "the whole nation should decide whether or not these experiments should continue." Common themes in both men's words are that Invit might be terribly deformed *à la thalidomide*, **resulting in a massive public backlash against all science,** and that the Edwards technique—if successful—might open the door to a *Brave New World* form of genetic engineering.

Edwards's administrative superiors in the physiology laboratory, however, defend the work. The Edwards artificial conception approach has been tested exhaustively in mammals, they say, and occurs in a stage of embryonic development when the danger of there being birth defects is at its lowest.

Interestingly, much of the impetus behind the Invit experiments stems from an unlikely source—the prospective mother.

Mother?

She will be one of some fifty willing women chosen by Edwards for the creation of Invit. The women are principally in their mid-thirties. Some are doctors; others are doctor's wives or nurses. They are sterile, principally because of blockages in their oviducts. Their ova, consequently, cannot make contact with sperm cells. They and their husbands above all want to become parents, and so they turned to Edwards when his experiments became public knowledge more than a year ago.

Edwards's scientific feat was to find exactly the right hormonal moment to remove eggs from a female volunteer (by laparoscopy, a technique that involves a needlelike instrument inserted through the navel into the egg sac), the right way to select sperm donated by the husband, and the correct liquid medium to encourage both sperm and egg to interact. He succeeded not only in achieving fertilization of an egg by such means but in coaxing the embryo to divide more than 100 times. This is more than enough to prepare Invit for "implantation," or attachment to the uterine wall—the moment many scientists believe to be the time when life really begins.

Edwards plans to induce the creation of Invit by bringing about contact of egg and sperm in an ordinary cell-culture dish, putting the budding mixture into an ordinary laboratory incubator, and then implanting the embryo at the appropriate stage of division—into the woman's receptive uterus, again by laparoscope. The mother-to-be will then go through the usual nine months of waiting as would any pregnant woman.

Invit, then, will be born. If normal, Invit will be examined, studied, interviewed, coddled, and analyzed for the rest of its life—but discreetly, in the Cambridge way. Invit will be shielded from a curious world, perhaps not even told of his (or her) origin.

And nature will have been bypassed in her most intimate and awesome of acts, conception.

203

ETHICAL PERSPECTIVES IN THE USE OF GENETIC KNOWLEDGE

Walter G. Peter III

"Ours is a time of intense self-doubt, corroding confidence, and crippling resolve; a time of troubled present and ominous future . . . and hence it is not surprising that so great a triumph as man's discovery of the molecular basis of inheritance should provoke fear instead of joy, breed suspicion instead of zest, and spawn the troubled anguish of indecision instead of the proud relief of understanding." With these words, Robert Sinsheimer, California Institute of Technology, expressed the growing anxiety with which scientists and laymen are viewing the increasing potentialities of the "new biology." It is this anxiety which brought 80 internationally renowned scholars to a secluded 4-day conference in the Washington, D.C. suburb of Warrenton, Virginia.

Jointly sponsored by the Institute of Society, Ethics and the Life Sciences and the John E. Fogarty International Center for Advanced Study in Health Sciences (NIH); the subject under discussion was "Ethical Issues in Genetic Counseling and the Use of Genetic Knowledge." It has become routine for observers of like meetings to write in summation that "meaningful dialogue was begun from which it is hoped that fruitful results will follow," or that "the necessary groundwork was laid from which further discussions can now begin." These stock refrains are usually an indication that a meeting was a disaster but the motives were noble. In the nonperjorative sense, both of these expressions accurately summarize this conference. Meaningful dialogue *was* begun and necessary groundwork *was* laid for further discussions. It remains to be seen if such a large conference can achieve greater goals.

Although the discussions were quite far-reaching, there seemed to be a constant reminder that the uses to which genetic knowledge will be put will not be decided by those who best understand the "new biology," but by those who understand it the least, the public or their appointed and elected representatives. The jurists and physicians who attended the conference often took the role of Public Advocate, although from different and sometimes conflicting positions, while the geneticists pondered the more elusive questions of theory, probability, and potentiality. It would be unfair, however, not to mention that several "cross-overs" took place whereby unlikely allies found themselves in the same camp. Genetic counseling and recent advances made in the detection of defective fetuses, primarily amniocentesis, brought into focus many ethi-

BIOSCIENCE, 1971, Vol. 21, pp. 1133-1137.

cal and legal questions. For example, geneticist John Littlefield, in his presentation "Prenatal Genetic Diagnosis: Status and Problems," stated that if a couple seeks a prenatal diagnosis of an inherited genetic disease, one presupposes that such a couple will elect therapeutic abortion if such disease is discovered. Although "medically indicated," this presupposition disturbed those who felt that therapeutic abortions could not be considered an "automatic" solution.

As miniaturized kinetic and enzyme assays become available, prenatal diagnosis will become shorter and the possibility of aborting defective fetuses greater. Perhaps, as Littlefield speculates, it may become possible to diagnose *in utero* serious autosomal dominant conditions such as neurofibromatosis, tuberous sclerosis, retinoblastoma, and Huntington's chorea. Ultimately, screening for all chromosomal and metabolic disorders may be done from amniotic fluid cells, thus making routine the monitoring of all pregnancies.

Daniel Callahan, philosopher and director of the Institute of Society Ethics and the Life Sciences, sees a potential danger to the right of self-determination should such mass screening become "routine" and abortion "automatic." He pointed out that until recently there was no "causal logic" discernible in bearing a defective child, therefore no burden of choice fell upon the unfortunate parents. If people are presented with a "freedom of choice" through the advances in genetics, then "giving people freedom of choice is to make them responsible for the choices they make. It is then only a very short step to begin distinguishing between responsible and irresponsible choices, with social pressure beginning to put in an appearance. Thus while, in principle, the parents of a fetus with a detected case of Down's Syndrome are still left free to decide whether to carry it to term, it is not difficult to discern an undercurrent in counseling literature and discussion that would classify such a decision as irresponsible. This is amplified in a subtle way. Abortion is said to be 'medically indicated' in such cases, as if what is essentially an ethical decision has now become nothing but medical." Callahan made two further points along this line. Along with the basic "right to life" issue is the one of financial costs. When one throws the "cost-benefit analysis" into the "genetic equation," we can now put a "price on everyone's head." He further stated, ". . . behind the human horror at genetic defectiveness lurks, one must suppose, an image of the perfect human being. The very language of 'defect,' 'abnormality,' 'disease,' and 'risk' presupposes such an image, a kind of prototype of perfection." This last point poses obvious questions of the notions of perfection and imperfection with all the political, social, and economic overtones.

Callahan did not reject the further development of genetic knowledge or the art of genetic counseling, but questioned the "spirit in which such an effort is taken . . . the kind of philosophical perspectives which lie behind it, and . . . the social context in which it is carried out."

These concerns brought out the fundamental issues of the rights of individuals vis-a-vis the rights of society. On a philosophical basis, Professor Aiken of Brandeis University saw at the heart of this issue the very definitions of "life" and "rights." He viewed the term "human being" as a normative concept based upon a special hierarchy of rights. He defined a "right" as an "expectation" plus "responsibility" or "obligation" in regard to that which is "expected." He defined "man" as simply a species name which represents a certain order of biological life. Therefore,

Aiken saw the question of "right to life" as "right to life as a human being" and "right to life as man." In the first instance, biological life must be present but, in the second instance, biological life could exist without the qualities of humanness. Thus, "a newborn baby is not as it stands a human being . . . it has the capacity if circumstances are favorable of becoming one." The implications of this stance are interesting in determining just when a newborn achieves the status of human being and when an aged human being, approaching a state of senility, loses his "humanness." Does Professor Aiken accept the possibility of infanticide and/or the selective culling of the species on the basis of rights lost? In fairness, Aiken was not given the time to substantiate fully his stands nor did he have the opportunity to explore these possible extrapolations. He admitted to overemphasis in order to make several contentions in a brief period of time.

The Practical Effects of Genetic Knowledge

The jurists had difficulty in differentiating between the concrete physiological and the abstract meanings of "life." As discussed by Blair L. Sadler in his paper, "The Law and the Unborn Child: A Brief Review of Emerging Problems," the fetus has been accorded the rights of a person in American Courts. For instance, the law provides for recovery by a deformed child if someone negligently hurt the mother while pregnant. Although there are some exceptions, "there is a distinct trend in the property, criminal and tort laws to regard the fetus as a human being from the time of conception and to accord the fetus rights consistent with this recognition." The courts have been steadily reversing anti-abortion laws while incurring inconsistencies with the above assertions. In

---

QUESTIONS DISCUSSED

Is a genetic counselor's responsibility first to the couple involved, or to society?

What constitutes a "defective" fetus?

When a mother being tested for Mongolism in her unborn child turns up negative in that respect, but the test shows the child has an extra "Y" chromosome—which *might* predispose him to antisocial behavior as an adult—should the doctor tell her? And if he does, what should she do?

Does an unborn baby have "rights" —including even the right not to be conceived (by cloning) as an identical twin of many others, or of his father? Can a parent ethically give "consent" on behalf of an unborn child—for gene manipulation, for example?

What will be the effect on society, the family, and the individual himself of being able to choose in advance which sex a child will be?

Mass prenatal screening can detect some 130 biochemical abnormalities. Who should do such screening? Which defects should have priority in the search? Should it be voluntary or compulsory?

Who should have access to the information from prenatal tests? How should it be used?

Could a child with genetic defects arising from a rubella infection early in pregnancy sue his mother's obstetrician—or his parents—for "wrongful life" in failing to abort him?

What public policy can and should be worked out to prevent abuses of current and future advances in genetics?

the Wisconsin case, *Babbitz v. McCann,* a three-judge Federal Court stated: "Upon a balancing of the relevant interests, we hold that a woman's right to refuse to carry an embryo during the early months of pregnancy may not be invaded by the state without a more compelling public necessity than is reflected in the statute in question. When measured against the claimed rights of the embryo of four months or less, we hold the mother's right transcends that of such an embryo." The operant words in these cases are "embryo" and "fetus" and the legal definition given to "human being."

Another case, *Geitman v. Cosgrove,* was described by Sadler and discussed by two other jurists at the Conference, Alex Capron of Yale and Lord Kilbrandon of the Scottish Law Commission. The case involved an infant who was born with serious birth defects after his mother had contracted German measles during her pregnancy. The infant and the mother both alleged that the defendant physicians were negligent in not informing the mother of the possible effects of German measles upon the infant, then in gestation. The State presented the infant plaintiff's case as follows: "But for the negligence of defendants, he would not have been born to suffer with an impaired body. In other words, he claims that the conduct of the defendants prevented his mother from obtaining an abortion which would have terminated his existence, and that his very life is 'wrongful'." Since tort law is based upon a doctrine of "compensatory damages," the court concluded that it "cannot weigh the value of life with impairments against the nonexistence of life itself."

Lord Kilbrandon supported the court in its decision and objects to those critics who reduce the argument to one where life without defects can be assigned a positive value (+), nonexistence a no-value (0), and life with defects a minus value (−). Kilbrandon states that "when an author attempts to express his diagram in words, he is forced into the use of an indefinite pronoun which conceals a semantic vacuum." "Such analysis," he (the author) says, "assumes that life without defects is to be desired most, but that in certain situations it would be preferable not to exist rather than to endure life incapacitated by severe physical and mental defects." Kilbrandon continues, "But for the words 'it would be preferable not to exist' you substitute as a plaintiff must do, the words 'I would prefer not to exist.' The fallacy is plain; the 'I' which is the subject being *ex hypothesi* nonexistent in one alternative cannot predicate his preference for another alternative. It would be like saying, 'I am glad I do not exist, because if I did, I would be mentally and physically handicapped.' This is non-language."

Alex Capron did not find fault with the decision concerning the infant plaintiff's suit, but he did find that the court was at fault for extending this line of thinking to the mother's and father's complaint. The court rejected the claims of mother and father who wished to be compensated for the emotional problems and expense of raising their deformed child. The court concluded, "The right of their child to live is greater than and precludes their right not to endure emotional and financial injury." It further contended that "even if such alleged damages were cognizable, a claim for them would be precluded by the countervailing public policy supporting the preciousness of human life." Capron took the position that the parents were denied two rights: the right of informed consent and, thusly, the right to self-determination. Ethically and legally, Capron contended that the physicians were at fault for not (1) explaining the effects of rubella upon the fetus, and (2) not explaining the alternatives open to the parents in regard to genetic counseling and possible abortion. The withholding of information by a physician is tantamount to denying a patient of "due process," and thus negligence.

Charles Fried, Professor of Law at Harvard, saw both Kilbrandon and Capron as missing the essential issue of developing a normative concept of human nature. While they delineated the immediate legal questions, they failed to realize that without a resolution of this essential problem, no solutions will be forthcoming. "The problem of rights, rights of individuals against each other, against the collectivity, even against the public good, is a modern problem. The modern liberal or individualistic concept of personality presupposes a concept of rights which the individual can claim even against the socially defined good. This concept of right is heavily involved in the issues which concern us here. Even as to the special question of genetic counseling, we ask whether a person has the right to certain kinds of information, the right to have his case treated confidentially, the right to an abortion, or conversely the right to marry and have children, and from a different angle again the right to be born, once conceived. In other areas of the 'new biology' there has been talk of the right to die, the right to be genetically unique, the right to be the product of normal sexual intercourse and pregnancy.... What we are far from having is a comprehensive theory of rights, much less a unified theory which shows how rights and the notion of the public good are functions of some overall scheme." What is "good," what do we "want," how shall we "choose" the wants and values of future generations?

As pointed out by Sinsheimer, and made obvious by the above discussion, our empirical advances in the field of genetics have far outstripped our ethical resolves. Cloning, extra-uterine gestation, and in vitro fertilization are causing scholars in all disciplines to search for new definitions, or reaffirmations of old definitions concerning "Life," "Rights," and "Genetic Defect."

Of immediate concern, as indicated by the Gleitman case, and of great importance to the physicians present, were two pressing problems: deciding upon whom the burden of disease is greatest and how to reconcile professional ethical norms when in conflict with the ethics of laymen. In the first instance, the burden might be carried exclusively by the parents of a defective child when that child is without pain or stress. In the case of a mongoloid child, the parents may bear enormous emotional and psychological stress, yet the child may be happy and, within its limits, well adjusted. Jerome LeJuene, geneticist from the Institut de Progenese, Paris, France, questioned how a scientific judgment could be made that would differentiate this kind of child from one that "should live." If suffering is to be a criterion, how do we assign values to the least possible suffering in this case. The child does not suffer at all as far as we know; it is the parents who suffer. How do you quantify suffering, how do you decide what is unbearable? In the second instance, several philosophers and ethicists were quick to point out to the physicians that if indeed a difference existed between their personal ethical standards and those of their profession that they were leading an unconscionable double life. Considerable disagreement was generated, however, concerning the disclosure of information by physician to patient.

A physician is bound by oath not to "do harm." A patient has the right to self-determination and informed consent. These two binding ethics can easily clash. If the jurists were persuasive in theory, specific examples given by physicians showed the anguish that the rigid enforcement of a principle could produce. One example that brought out the dilemma was the case of a child produced by an extra-marital union unknown to the "parents." Such cases have occurred when genetic counseling has been sought and such illegitimacy is discovered. If the family is happy and well-adjusted, what are the obligations of the counselor? Another case presented involved a patient of one of the attending physicians. The patient inherited a rare anomaly known as testicular femination. Although possessing imperfect testicles in place of ovaries and XY chromosomes, she is quite definitely female in *all* other ways and happily married. The gonads must be removed since there is a high risk of cancer involved in such cases. Should the physician "tell the truth"? If the physician did reveal the truth, what of the possible psychological consequences? Could he be sued for "causing harm" if his patient were to become acutely depressed or if the marriage ended in divorce?

Is Mass Screening Desirable?

Another immediate reality of genetic knowledge is the use of mass screening techniques to identify the carriers of genetic traits for disease. In Baltimore and Washington, D.C., a highly successful screening was undertaken under the direction of Johns Hopkins University physician, Michael Kabak, for a very rare anomaly, Tay Sachs disease. Striking the child of parents who are heterozygous carriers of East European Jewish descent, the disease is 100% lethal within the first 6 years of life. It is neurodegenerating, progressive, and shows the symptoms of blindness, total paralysis, and total mental retardation. As far as can be determined, the child suffers no pain. The financial cost of maintaining a Tay Sachs baby is about $40.000 annually. The chances of a couple heterozygous for Tay Sachs of bearing a child with the disease is 1 in 4. It can be detected in a fetus during the early second trimester of pregnancy; therefore, the use of amniocentesis and, in the event of a positive diagnosis, therapeutic abortion offer a viable option to such carriers to safely have normal children each and every time.

Prior to the screening, a carefully planned educational program was undertaken so that the community to be screened was knowledgeable about the disease, the screening procedures, and confidentiality was ensured.

By contrast, there has been widespread pressure to screen for the sickle-cell trait—carried only by our Black population. Robert Murray, Chief of the Medical Genetics Unit of Howard University, sees such screening as playing upon the fears and anxieties of a carrier group for which there is no therapy available. A metabolic disease, sickle-cell anemia does not follow Mendel's law of probability. As Murray put it, "What do you tell a Black man or woman who carries the sickle-cell trait?" Outside of the possible stigma attached to being different, hardly a needed further detriment to a Black person, he saw no value in mass screening for this disease.

Murray further stated that the incidence of sickle cell anemia is so low that it hardly ranks with the other problems facing the Black man. He sees the mass screening of this population as symbolically very important to those who advocate it and, realistically, inconsequential to Blacks. David Eaton, a Unitarian Minister and well-known community leader in Washington, D.C., agreed with Murray and stated unequivocally that he viewed such measures as further evidence of the White man's paternalism. He said that he and most Blacks were fighting for survival by combating malnutrition, drug abuse, communicable disease, and a host of other poverty-related ills of far greater consequence than genetic disease. He queried, "How about trying to do something to save healthy fetuses?"

What Priorities Should Exist?

At the end of the Conference, James V. Neel, University of Michigan, presented a list of priorities in the use of scientific genetic knowledge with the following introduction: ". . . I am sure we would agree that for the next 10-20 years, the impact of failure to apply the new genetic knowledge to human problems is far less than failure to solve such major issues as control of pollution, decay of cities, or a sane energy policy. On the other hand, as I believe has been apparent from previous presentations, the potential of genetic knowledge for the good of man is really very great. Furthermore, there is a symbolism about the use of genetic knowledge to influence the genetic composition of the next generation which readily captures the public imagination. Herein lies one of our problems in setting priorities. Right now we are something of a glamour field. . . . We do have important contributions to make to human well-being. However, we have caught the public imagination, and it's going to take a great deal of intellectual honesty and sobriety not to take advantage of this fact."

Neel then presented the following list of priorities: (1) genetic counseling and its logical extension, prenatal diagnosis; (2) genetic screening; (3) genetic repair, either somatic or germinal; (4) prevention of mutation; (5) amelioration of genotype expression; (6) germinal choice; and (7) stabilization of the gene pool. The advances in these areas could be measured against the following criteria: (1) the reduction of the propor-

tion of persons with genetic disease, an objective with which we most readily associate genetic counseling, prenatal diagnosis, and, perhaps, genetic surgery; (2) the creation of genetically superior individuals, by artificial insemination, perhaps cloning; (3) the improvement of the expression of existing genotypes by wide-ranging medical, social, and nutritional methods; (4) the protection of the present gene pool by a world population policy which will at least ensure that as little as possible of what now exists is lost, and damage through exposure to mutagens is minimized; (5) a minimum of incalculable genetic and somatic risks.

Neel asked "let the record show that before we began our discussion of the difficult issue of setting priorities in this field, we recognized that some might consider this discussion both presumptuous and naive. Our discussion should deserve neither of these adjectives as long as we recognize this as an exercise in social and scientific judgment, approached with the necessary humility and objectivity."

There are many questions that were not examined. It is a well-known biological principle that heterogenity and variety are insurors of long species survival. Should eugenics be used to maximize specific limited characteristics, such as might occur with cloning, would this not substantially lower our survival chances? As population control through the limitation of progeny becomes reality, prospective parents will be more concerned about the quality of their children. Will they demand the "right" of genetic counseling? How will we decide which defects constitute reason for abortion and which do not? Will future genetic knowledge alter our understanding of the term "disease"? Will what is now considered a mild or inconsequential anomaly become an undesired defect? Who will make these determinations?

These are but a few examples of issues that will be forthcoming if genetic knowledge continues to proceed at its present pace. There is some question in the minds of geneticists as to whether such research should even proceed or, if it does, whether it should ever be made public. In the words of Sinsheimer, "Much of the despair of our time stems from the realization that—at last—after all the toil and all the inventions and all the savagery and all the genius—the enemy is 'us.' Our deepest problems are now 'man made.' Their origin lies in the inherent corrosion of imperfect man. Should we arm this creature with vast new powers?"

Bibliography

1. Augenstein, L. G. 1968. Come, Let Us Play God. Harper and Row, New York.

2. Fuhrmann, Walter, and Friedrich Vogel. 1969. Genetic Counseling. Springer-Verlag, New York.

3. Grobman, A. B. (ed.). 1970. Social Implications of Biological Education. National Association of Biology Teachers, Washington, D.C.

4. Hamilton, Michael (ed.) 1972. The New Genetics and the Future of Man. Wm. B. Eerdmans, Grand Rapids, Michigan.

5. Lerner, I. M. 1968. Heredity, Evolution and Society. W. H. Freeman, San Francisco.

6. Levitan, Max and Ashley Montagu. 1971. Textbook of Human Genetics. Oxford University Press, New York.

7. McKusick, V. A. 1969. Human Genetics, Second Edition. Prentice-Hall, Englewood Cliffs, New Jersey.

8. Moore, J. A. 1971. Science for Society: A Bibliography, Second Edition. Commission on Science Education. American Association for the Advancement of Science, Washington, D. C.

9. Ramsey, Paul. 1970. Fabricated Man - The Ethics of Genetic Control. Yale University Press, New Haven, Conn.

10. Roslansky, J. D. (ed.). 1966. Genetics and the Future of Man. Appleton - Century - Crofts, New York.

11. Sonneborn, T. M. (ed.). 1965. The Control of Human Heredity and Evolution. Macmillan, New York.

12. Taylor, G. R. 1968. The Biological Time Bomb. World, New York.

13. Young, L. B. (ed.). 1970. Evolution of Man. Oxford University Press, New York.